U.S. ARMY LEADERSHIP HANDBOOK

Skills, Tactics, and Techniques for Leading in Any Situation

DEPARTMENT OF THE ARMY

Skyhorse Publishing

Copyright © 2012 by Skyhorse Publishing, Inc.

No claim is made to material contained in this work that is derived from government documents. Nevertheless, Skyhorse Publishing, claims copyright in all additional content, including, but not limited to, compilation copyright and the copyright in and to any additional material, elements, design, or layout of whatever kind included herein.

All Rights Reserved. No part of this book may be reproduced in any manner without the express written consent of the publisher, except in the case of brief excerpts in critical reviews or articles. All inquiries should be addressed to Skyhorse Publishing, 307 West 36th Street, 11th Floor, New York, NY 10018.

Skyhorse Publishing books may be purchased in bulk at special discounts for sales promotion, corporate gifts, fund-raising, or educational purposes. Special editions can also be created to specifications. For details, contact the Special Sales Department, Skyhorse Publishing, 307 West 36th Street, 11th Floor, New York, NY 10018 or info@skyhorsepublishing.com.

Skyhorse® and Skyhorse Publishing® are registered trademarks of Skyhorse Publishing, Inc.®, a Delaware corporation.

Visit our website at www.skyhorsepublishing.com.

10 9 8

Library of Congress Cataloging-in-Publication Data

United States. Dept. of the Army.
U.S. Army leadership handbook : skills, tactics, and techniques for leading in any situation / Department of the Army.
p. cm.
Includes index.
ISBN 978-1-61608-562-9 (alk. paper)
1. United States. Army--Management--Handbooks, manuals, etc. 2. Military education--Handbooks, manuals, etc. 3. Armed Forces--Officers--Handbooks, manuals, etc. 4. Leadership--Handbooks, manuals, etc. 5. Command of troops--Handbooks, manuals, etc. I. Title.
UB210.U517 2012
355.3'3041--dc23

2012001670

Printed in China

***FM 6-22 (FM 22-100)**

Field Manual
No. 6-22

Headquarters
Department of the Army
Washington, DC

U.S. Army Leadership Handbook

Contents

Distribution Restriction: Approved for public release; distribution is unlimited.

This publication supersedes FM 22-100, 31 August 1999.

Figures

Foreword

Competent leaders of character are necessary for the Army to meet the challenges in the dangerous and complex security environment we face.

FM 6-22 is the Army's keystone field manual on leadership. It establishes leadership doctrine and fundamental principles for all officers, noncommissioned officers, and Army civilians across all components.

This manual uses the BE-KNOW-DO concept to express what is required of Army leaders. It is critical that Army leaders be agile, multiskilled *pentathletes* who have strong moral character, broad knowledge, and keen intellect. They must display these attributes and leader competencies bound by the concept of the Warrior Ethos. Leaders must be committed to lifelong learning to remain relevant and ready during a career of service to the Nation.

Army leaders must set the example, teach, and mentor, and this manual provides the principles, concepts, and training to accomplish this important task on which America depends.

PETER J. SCHOOMAKER
General, United States Army
Chief of Staff

This publication is available at Army Knowledge Online (www.us.army.mil) and General Dennis J. Reimer Training and Doctrine Digital Library at (www.train.army.mil).

Preface

As the keystone leadership manual for the United States Army, FM 6-22 establishes leadership doctrine, the fundamental principles by which Army leaders act to accomplish their mission and care for their people. FM 6-22 applies to officers, warrant officers, noncommissioned officers, and enlisted Soldiers of all Army components, and to Army civilians. From Soldiers in basic training to newly commissioned officers, new leaders learn how to lead with this manual as a basis.

FM 6-22 is prepared under the direction of the Army Chief of Staff. It defines leadership, leadership roles and requirements, and how to develop leadership within the Army. It outlines the levels of leadership as direct, organizational, and strategic, and describes how to lead successfully at each level. It establishes and describes the core leader competencies that facilitate focused feedback, education, training, and development across all leadership levels. It reiterates the Army Values. FM 6-22 defines how the Warrior Ethos is an integral part of every Soldier's life. It incorporates the leadership qualities of self-awareness and adaptability and describes their critical impact on acquiring additional knowledge and improving in the core leader competencies while operating in constantly changing operational environments.

In line with evolving Army doctrine, FM 6-22 directly supports the Army's capstone manuals, FM 1 and FM 3-0, as well as keystone manuals such as FM 5-0, FM 6-0, and FM 7-0. FM 6-22 connects Army doctrine to joint doctrine as expressed in the relevant joint doctrinal publications, JP 1 and JP 3-0.

As outlined in FM 1, the Army uses the shorthand expression of BE-KNOW-DO to concentrate on key factors of leadership. What leaders DO emerges from who they are (BE) and what they KNOW. Leaders are prepared throughout their lifetimes with respect to BE-KNOW-DO so they will be able to act at a moment's notice and provide leadership for whatever challenge they may face.

FM 6-22 expands on the principles in FM 1 and describes the character attributes and core competencies required of contemporary leaders. Character is based on the attributes central to a leader's make-up, and competence comes from how character combines with knowledge, skills, and behaviors to result in leadership. Inextricably linked to the inherent qualities of the Army leader, the concept of BE-KNOW-DO represents specified elements of character, knowledge, and behavior described here in FM 6-22.

This publication contains copyrighted material.

This publication applies to all men and women of all ranks and grades who serve in the Active Army, Army National Guard/Army National Guard of the United States, Army Reserve, and Army civilian workforce unless otherwise stated.

Terms that have joint or Army definitions are identified in both the glossary and the text. *Glossary references*: Terms for which FM 6-22 is the proponent field manual (the authority) are indicated with an asterisk in the glossary. *Text references*: Definitions for which FM 6-22 is the proponent field manual are printed in boldface in the text. These terms and their definitions will be incorporated into the next revision of FM 1-02. For other definitions in the text, the term is italicized and the number of the proponent manual follows the definition.

U.S. Army Training and Doctrine Command is the proponent for this publication. The preparing agency is the Center for Army Leadership, Command and General Staff College. Send written comments and recommendations on DA Form 2028 (Recommended Changes to Publications and Blank Forms) to Center for Army Leadership ATTN: ATZL-CAL (FM 6-22), 250 Gibbon Avenue, Fort Leavenworth, KS 66027-2337. Send comments and recommendations by e-mail to ***calfm622@leavenworth.army.mil***. Follow the DA Form 2028 format or submit an electronic DA Form 2028.

Acknowledgements

The copyright owners listed here have granted permission to reproduce material from their works. Other sources of quotations and material used in examples are listed in the source notes.

The example Colonel Chamberlain at Gettysburg is adapted from John J. Pullen, *The Twentieth Maine* (1957; reprint, Dayton, OH, Press of Morningside Bookshop, 1980). Reprinted with permission of the Estate of John Pullen via The Ward & Balkin Agency, Inc.

The section on Vertical Command Teams in Chapter 3 is reproduced from IDA Document D-2728 by LTG (ret.) Frederic J. Brown "Vertical Command Teams" (Alexandria, VA: Institute for Defense Analysis, 2002).

The quotation by Douglas E. Murray in Chapter 3 is reproduced from Dennis Steele, "Broadening the Picture Calls for Turning Leadership Styles," *Army Magazine*, December 1989. Copyright © 1989 by the Association of the United States Army and reproduced by permission.

The example Shared Leadership Solves Logistics Challenges in Chapter 3 is adapted from the website article, John Pike, "Operation Enduring Freedom-Afghanistan" (http://www.globalsecurity.org, March 2005).

The quotation by William Connelly in Chapter 4 is reproduced from William Connelly, "NCOs: It's Time to Get Tough," *Army Magazine*, October 1981. Copyright © 1981 by the Association of the United States Army and reproduced by permission.

The example Task Force Kingston is adapted from Martin Blumenson, "Task Force Kingston," *Army Magazine*, April 1964. Copyright © 1964 by the Association of the United States Army and reproduced by permission.

The section on Ethical Reasoning in Chapter 4 is adapted from Michael Josephson, *Making Ethical Decisions* with permission of the Josephson Institute of Ethics. Copyright © 2002. www.charactercounts.org.

The example and quotation within Self Control in Chapter 6 is adapted from *Leader to Leader*, ed. Francis Hesselbein (New York: Leader to Leader Institute, 2005). Copyright © (2005 Francis Hesselbein). Reproduced with permission of John Wiley & Sons Inc.

The section on The Quickest and Most Efficient Way to Plan in chapter 12 is reproduced from MSG (ret.) Paul R. Howe, *Leadership and Training for the Fight*, 170–172 (New York: Skyhorse Publishing, 2011). Copyright © 2011 by Paul R. Howe and reproduced by permission.

The quotation by Major Richard Winters in Chapter 7 is reproduced from Christopher J. Anderson, "Dick Winters" Reflections on His Band of Brothers, D-Day, and Leadership," *American History Magazine* (August 2004). Reproduced with permission of *American History Magazine*, Primedia Enthusiast Publications.

The quotation by Richard A. Kidd in Chapter 8 is reproduced from Richard A. Kidd, "NCOs Make it Happen," *Army Magazine*, October 1994. Copyright © 1994 by the Association of the United States Army and reproduced by permission.

The quotation by William Connelly in Chapter 8 is reproduced from William A. Connelly, "Keep Up with Change in the 80's," *Army Magazine*, October 1976. Copyright © 1976 by the Association of the United States Army and reproduced by permission.

The quotation in Chapter 9 by William C. Bainbridge is reproduced from William C. Bainbridge, "Quality, Training and Motivation," *Army Magazine*, October 1976. Copyright © 1994 by the Association of the United States Army and reproduced by permission.

Introduction

Upon taking the oath to become an Army leader, Soldiers, and Army civilians enter into a sacred agreement with the Nation and their subordinates. The men and women of the Army are capable of extraordinary feats of courage and sacrifice as they have proven on countless battlefields from the Revolutionary War to the War on Terrorism. These Soldiers and Army civilians display great patience, persistence, and tremendous loyalty as they perform their duty to the Nation in thousands of orderly rooms, offices, motor pools, and training areas around the world, no matter how difficult, tedious, or risky the task. In return, they deserve competent, professional, and ethical leadership. They expect their Army leaders to respect them as valued members of effective and cohesive organizations and to embrace the essence of leadership.

FM 6-22 combines the lessons of the past with important insights for the future to help develop competent Army leaders.

An ideal Army leader has strong intellect, physical presence, professional competence, high moral character, and serves as a role model. An Army leader is able and willing to act decisively, within the intent and purpose of his superior leaders, and in the best interest of the organization. Army leaders recognize that organizations, built on mutual trust and confidence, successfully accomplish peacetime and wartime missions.

Organizations have many leaders. Everyone in the Army is part of a chain of command and functions in the role of leader and subordinate. Being a good subordinate is part of being a good leader. All Soldiers and Army civilians, at one time or another, must act as leaders and followers. Leaders are not always designated by position, rank, or authority. In many situations, it is appropriate for an individual to step forward and assume the role of leader. It is important to understand that leaders do not just lead subordinates—they also lead other leaders.

Everyone in the Army is part of a team, and all team members have responsibilities inherent in belonging to that team.

FM 6-22 addresses the following topics necessary to become a competent, multiskilled Army leader:

- Understand the Army definitions of leader and leadership.
- Learn how the Warrior Ethos is embedded in all aspects of leadership.
- Use the Army leadership requirements model as a common basis for thinking and learning about leadership and associated doctrine.
- Become knowledgeable about the roles and relationships of leaders, including the role of subordinate or team member.
- Discover what makes a good leader, a person of character with presence and intellect.
- Learn how to lead, develop, and achieve through competency-based leadership.
- Identify the influences and stresses in our changing environment that affect leadership.
- Understand the basics of operating at the direct, organizational, and strategic levels.

PART ONE

The Basis of Leadership

All Army team members, Soldiers and civilians alike, must have a basis of understanding for what leadership is and does. The definitions of leadership and leaders address their sources of strength in deep-rooted values, the Warrior Ethos, and professional competence. National and Army values influence the leader's character and professional development, instilling a desire to acquire the essential knowledge to lead. Leaders apply this knowledge within a spectrum of established competencies to achieve successful mission accomplishment. The roles and functions of Army leaders apply to the three interconnected levels of leadership: direct, organizational, and strategic. Within these levels of leadership, cohesive teams can achieve collective excellence when leadership levels interact effectively.

Chapter 1

Leadership Defined

1-1. An enduring expression for Army leadership has been BE-KNOW-DO. Army leadership begins with what the leader must **BE**—the values and attributes that shape character. It may be helpful to think of these as internal and defining qualities possessed all the time. As defining qualities, they make up the identity of the leader.

1-2. Who is an Army leader?

An ***Army leader*** is anyone who by virtue of assumed role or assigned responsibility inspires and influences people to accomplish organizational goals. Army leaders motivate people both inside and outside the chain of command to pursue actions, focus thinking, and shape decisions for the greater good of the organization.

1-3. Values and attributes are the same for all leaders, regardless of position, although refined through experience and assumption of positions of greater responsibility. For example, a sergeant major with combat experience may have a deeper understanding of selfless service and personal courage than a new Soldier.

1-4. The knowledge that leaders should use in leadership is what Soldiers and Army civilians **KNOW**. Leadership requires knowing about tactics, technical systems, organizations, management of resources, and the tendencies and needs of people. Knowledge shapes a leader's identity and is reinforced by a leader's actions.

1-5. While character and knowledge are necessary, by themselves they are not enough. Leaders cannot be effective until they apply what they know. What leaders **DO,** or leader actions, is directly related to the influence they have on others and what is done. As with knowledge, leaders will learn more about leadership as they serve in different positions.

1-6. New challenges facing leaders, the Army, and the Nation mandate adjustments in how the Army educates, trains, and develops its military and civilian leadership. The Army's mission is to fight and win the

Nation's wars by providing prompt, sustained land dominance across the spectrum of conflicts in support of combatant commanders. In a sense, all Army leaders must be warriors, regardless of service, branch, gender, status, or component. All serve for the common purpose of protecting the Nation and accomplishing their organization's mission to that end. They do this through influencing people and providing purpose, direction, and motivation.

Leadership is the process of influencing people by providing purpose, direction, and motivation while operating to accomplish the mission and improving the organization.

INFLUENCING

1-7. Influencing is getting people—Soldiers, Army civilians, and multinational partners—to do what is necessary. Influencing entails more than simply passing along orders. Personal examples are as important as spoken words. Leaders set that example, good or bad, with every action taken and word spoken, on or off duty. Through words and personal example, leaders communicate purpose, direction, and motivation.

PURPOSE AND VISION

1-8. Purpose gives subordinates the reason to act in order to achieve a desired outcome. Leaders should provide clear purpose for their followers and do that in a variety of ways. Leaders can use direct means of conveying purpose through requests or orders for what to do.

1-9. Vision is another way that leaders can provide purpose. Vision refers to an organizational purpose that may be broader or have less immediate consequences than other purpose statements. Higher-level leaders carefully consider how to communicate their vision.

DIRECTION

1-10. Providing clear direction involves communicating how to accomplish a mission: prioritizing tasks, assigning responsibility for completion, and ensuring subordinates understand the standard. Although subordinates want and need direction, they expect challenging tasks, quality training, and adequate resources. They should be given appropriate freedom of action. Providing clear direction allows followers the freedom to modify plans and orders to adapt to changing circumstances. Directing while adapting to change is a continuous process.

1-11. For example, a battalion motor sergeant always takes the time and has the patience to explain to the mechanics what is required of them. The sergeant does it by calling them together for a few minutes to talk about the workload and the time constraints. Although many Soldiers tire of hearing from the sergeant about how well they are doing and that they are essential to mission accomplishment, they know it is true and appreciate the comments. Every time the motor sergeant passes information during a meeting, he sends a clear signal: people are cared for and valued. The payoff ultimately comes when the unit is alerted for a combat deployment. As events unfold at breakneck speed, the motor sergeant will not have time to explain, acknowledge performance, or motivate them. Soldiers will do their jobs because their leader has earned their trust.

MOTIVATION

1-12. Motivation supplies the will to do what is necessary to accomplish a mission. Motivation comes from within, but is affected by others' actions and words. A leader's role in motivation is to understand the needs and desires of others, to align and elevate individual drives into team goals, and to influence others and accomplish those larger aims. Some people have high levels of internal motivation to get a job done, while others need more reassurance and feedback. Motivation spurs initiative when something needs to be accomplished.

1-13. Soldiers and Army civilians become members of the Army team for the challenge. That is why it is important to keep them motivated with demanding assignments and missions. As a leader, learn as much as possible about others' capabilities and limitations, then give over as much responsibility as can be handled.

When subordinates succeed, praise them. When they fall short, give them credit for what they have done right, but advise them on how to do better. When motivating with words, leaders should use more than just empty phrases; they should personalize the message.

1-14. Indirect approaches can be as successful as what is said. Setting a personal example can sustain the drive in others. This becomes apparent when leaders share the hardships. When a unit prepares for an emergency deployment, all key leaders should be involved to share in the hard work to get the equipment ready to ship. This includes leadership presence at night, weekends, and in all locations and conditions where the troops are toiling.

OPERATING

1-15. Operating encompasses the actions taken to influence others to accomplish missions and to set the stage for future operations. One example is the motor sergeant who ensures that vehicles roll out on time and that they are combat ready. The sergeant does it through planning and preparing (laying out the work and making necessary arrangements), executing (doing the job), and assessing (learning how to work smarter next time). The motor sergeant leads by personal example to achieve mission accomplishment. The civilian supervisor of training developers follows the same sort of operating actions. All leaders execute these types of actions which become more complex as they assume positions of increasing responsibility.

IMPROVING

1-16. Improving for the future means capturing and acting on important lessons of ongoing and completed projects and missions. After checking to ensure that all tools are repaired, cleaned, accounted for, and properly stowed away, our motor sergeant conducts an after-action review (AAR). An AAR is a professional discussion of an event, focused on performance standards. It allows participants to discover for themselves what happened, why it happened, how to sustain strengths, and how to improve on weaknesses. Capitalizing on honest feedback, the motor sergeant identifies strong areas to sustain and weak areas to improve. If the AAR identifies that team members spent too much time on certain tasks while neglecting others, the leader might improve the section standing operating procedures or counsel specific people on how to do better.

1-17. Developmental counseling is crucial for helping subordinates improve performance and prepare for future responsibilities. The counseling should address strong areas as well as weak ones. If the motor sergeant discovers recurring deficiencies in individual or collective skills, remedial training is planned and conducted to improve these specific performance areas. Part Three and Appendix B provide more information on counseling.

1-18. By stressing the team effort and focused learning, the motor sergeant gradually and continuously improves the unit. The sergeant's personal example sends an important message to the entire team: Improving the organization is everyone's responsibility. The team effort to do something about its shortcomings is more powerful than any lecture.

Chapter 2

The Foundations of Army Leadership

2-1. The foundations of Army leadership are firmly grounded in history, loyalty to our country's laws, accountability to authority, and evolving Army doctrine. By applying this knowledge with confidence and dedication, leaders develop into mature, competent, and multiskilled members of the Nation's Army. While Army leaders are responsible for being personally and professionally competent, they are also charged with the responsibility of developing their subordinates.

2-2. To assist leaders to become competent at all levels of leadership, the Army identifies three categories of core leader competencies: lead, develop, and achieve. These competencies and their subsets represent the roles and functions of leaders.

THE FOUNDING DOCUMENTS OF OUR NATION

When we assumed the Soldier, we did not lay aside the Citizen.

General George Washington
Speech to the New York Legislature, 1775

2-3. The Army and its leadership requirements are based on the Nation's democratic foundations, defined values, and standards of excellence. The Army recognizes the importance of preserving the time-proven standards of competence that have distinguished leaders throughout history. Leadership doctrine acknowledges that societal change, evolving security threats, and technological advances require an ever-increasing degree of adaptability.

2-4. Although America's history and cultural traditions derive from many parts of the civilized world, common values, goals, and beliefs are solidly established in the Declaration of Independence and the Constitution. These documents explain the purpose of our nationhood and detail our specific freedoms and responsibilities. Every Army Soldier and leader should be familiar with these documents.

2-5. On 4 July 1776, the Declaration of Independence formally sealed America's separation from British rule and asserted her right as an equal participant in dealings with other sovereign nations. Adopted by Congress in March of 1787, the U.S. Constitution formally established the basic functions of our democratic government. It clearly explains the functions, as well as the checks and balances between the three branches of government: the executive, the legislative, and the judicial. The Constitution sets the parameters for the creation of our national defense establishment, including the legal basis for our Army. Amended to the Constitution in December 1791, the Federal Bill of Rights officially recognized specific rights for every American citizen, including freedom of religion, of speech, and of the press. At the time of publication of FM 6-22, there have been 27 amendments to the Constitution. The amendments illustrate the adaptability of our form of government to societal changes.

THE CIVILIAN-MILITARY LINKAGE

2-6. The U.S. Constitution grants Congress the ability to raise and support armies. Subsequently, the armed forces are given the task of defending the United States of America and her territories. Membership in the Army and its other Services is marked by a special status in law. That status is reflected in distinctive uniforms and insignia of service and authority. To be able to function effectively on the battlefield, the Army and other Services are organized into hierarchies of authority. The Army's hierarchy begins with the individual Soldier and extends through the ranks to the civilian leadership including the Secretary of the Army, Secretary of Defense, and the President of the United States.

2-7. To formalize our ties to the Nation and to affirm subordination to its laws, members of the Army—Soldiers and Army civilians—swear a solemn oath to support and defend the Constitution of the United States against all enemies, foreign and domestic. Soldiers simultaneously acknowledge the authority of the President as Commander in Chief and officers as his agents. The purpose of the oath is to affirm military subordination to civilian authority. The Army Values in figure 2-1 link tightly with the content of the oath.

> *I do solemnly swear (or affirm) that I will support and defend the Constitution of the United States against all enemies, foreign and domestic; that I will bear true faith and allegiance to the same; and that I will obey the orders of the President of the United States and the orders of the officers appointed over me, according to regulations and the Uniform Code of Military Justice. So help me God.*
>
> Oath of Enlistment

> *I do solemnly swear (or affirm) that I will support and defend the Constitution of the United States against all enemies, foreign and domestic; that I will bear true faith and allegiance to the same; that I take this obligation freely, without any mental reservation or purpose of evasion; and that I will well and faithfully discharge the duties of the office on which I am about to enter. So help me God.*
>
> Oath of office taken by commissioned officers and Army civilians

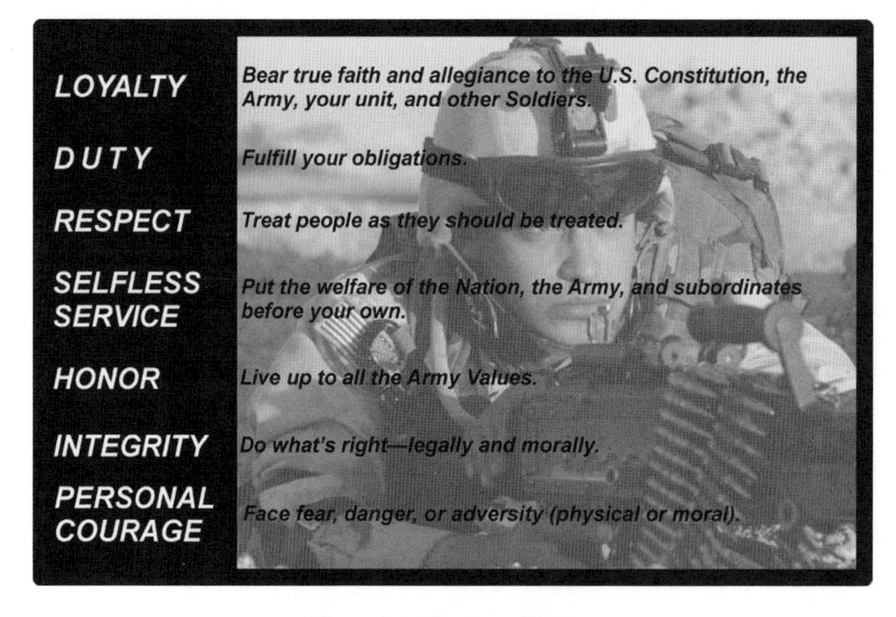

Figure 2-1. The Army Values

2-8. The oath and values emphasize that the Army's military and civilian leaders are instruments of the people of the United States. The elected government commits forces only after due consideration and in compliance with our national laws and values. Understanding this process gives our Army moral strength and unwavering confidence when committed to war.

2-9. As General George Washington expressed more than 200 years ago, serving as a Soldier of the United States does not mean giving up being an American citizen with its inherent rights and responsibilities. Soldiers are citizens and should recognize that when in uniform, they represent their units, their Army, and their country. Every Soldier must balance the functions of being a dedicated warrior with obedience to the laws of the Nation. They must function as ambassadors for the country in peace and war. Similarly, self-disciplined behavior is expected of Army civilians.

LEADERSHIP AND COMMAND AUTHORITY

When you are commanding, leading [Soldiers] under conditions where physical exhaustion and privations must be ignored; where the lives of [Soldiers] may be sacrificed, then, the efficiency of your leadership will depend only to a minor degree on your tactical or technical ability. It will primarily be determined by your character, your reputation, not so much for courage—which will be accepted as a matter of course—but by the previous reputation you have established for fairness, for that high-minded patriotic purpose, that quality of unswerving determination to carry through any military task assigned you.

General of the Army George C. Marshall
Speaking to officer candidates (1941)

2-10. Command is a specific and legal leadership responsibility unique to the military.

Command is the authority that a commander in the military service lawfully exercises over subordinates by virtue of rank or assignment. Command includes the leadership, authority, responsibility, and accountability for effectively using available resources and planning the employment of, organizing, directing, coordinating, and controlling military forces to accomplish assigned missions. It includes responsibility for unit readiness, health, welfare, morale, and discipline of assigned personnel (FMI 5-0.1).

2-11. Command is about sacred trust. Nowhere else do superiors have to answer for how their subordinates live and act beyond duty hours. Society and the Army look to commanders to ensure that Soldiers and Army civilians receive the proper training and care, uphold expected values, and accomplish assigned missions.

2-12. In Army organizations, commanders set the standards and policies for achieving and rewarding superior performance, as well as for punishing misconduct. In fact, military commanders can enforce their orders by force of criminal law. Consequently, it should not come as a surprise that organizations often take on the personality of their commanders. Army leaders selected to command are expected to lead beyond merely exercising formal authority. They should lead by example and serve as role models, since their personal example and public actions carry tremendous moral force. For that reason, people inside and outside the Army recognize commanders as the human faces of the system, the ones who embody the Army's commitment to readiness and care of people. By virtue of their role, Army commanders must lead change with clear vision, encompassing yesterday's heritage, today's mission, and tomorrow's force.

THE ARMY LEADERSHIP REQUIREMENTS MODEL

Just as the diamond requires three properties for its formation—carbon, heat, and pressure—successful leaders require the interaction of three properties—character, knowledge, and application. Like carbon to the diamond, character *is the basic quality of the leader.... But as carbon alone does not create a diamond, neither can character alone create a leader. The diamond needs heat. Man needs* knowledge, *study and preparation.... The third property, pressure—acting in conjunction with carbon and heat—forms the diamond. Similarly, one's character attended by knowledge, blooms through* application *to produce a leader.*

General Edward C. Meyer
Chief of Staff, Army (1979-1983)

2-13. FM 1, one of the Army's two capstone manuals, states that the Army exists to serve the American people, protect enduring national interests, and fulfill the Nation's military responsibilities. To accomplish this requires values-based leadership, impeccable character, and professional competence. Figure 2-2 shows the Army leadership requirements model. It provides a common basis for thinking and learning about leadership and associated doctrine. All of the model's components are interrelated.

Figure 2-2. The Army leadership requirements model

2-14. The model's basic components center on what a leader is and what a leader does. The leader's character, presence, and intellect enable the leader to master the core leader competencies through dedicated lifelong learning. The balanced application of the critical leadership requirements empowers the Army leader to build high-performing and cohesive organizations able to effectively project and support landpower. It also creates positive organizational climates, allowing for individual and team learning, and empathy for all team members, Soldiers, civilians, and their families.

2-15. Three major factors determine a leader's character: values, empathy, and the Warrior Ethos. Some characteristics are present at the beginning of the leader's career, while others develop over time through additional education, training, and experience.

2-16. A leader's physical presence determines how others perceive that leader. The factors of physical presence are military bearing, physical fitness, confidence, and resilience. The leader's intellectual capacity helps to conceptualize solutions and acquire knowledge to do the job. A leader's conceptual abilities apply agility, judgment, innovation, interpersonal tact, and domain knowledge. Domain knowledge encompasses tactical and technical knowledge as well as cultural and geopolitical awareness.

2-17. The famous fight between the 20th Regiment of Maine Volunteers and the 15th and 47th Regiments of Alabama Infantry during the battle of Gettysburg shows multiple components of the leadership requirements model at work. At the focal point, Colonel Joshua Chamberlain, a competent and confident leader, turns a seemingly hopeless situation into victory.

Colonel Chamberlain at Gettysburg

In late June 1863, Confederate GEN Robert E. Lee's Army of Northern Virginia passed through western Maryland and invaded Pennsylvania. For five days, the Army of the Potomac hurried to get between the Confederates and the National capital. On 1 July 1863, the 20th Maine received word to press on to Gettysburg. The Union Army had engaged the Confederates there, and Union commanders were hurrying all available forces to the hills south of the little town.

The 20th Maine arrived at Gettysburg near midday on 2 July, after marching more than one hundred miles in five days. They had had only two hours sleep and no hot food during the previous 24 hours. The regiment was preparing to go into a defensive position as part of the brigade commanded by COL Strong Vincent when a staff officer rode up to COL Vincent and began gesturing towards a little hill at the extreme southern end of the Union line. The hill, Little Round Top, dominated the Union position and, at that moment, was unoccupied. If the Confederates placed artillery on it, they could force the entire Union Army to withdraw. The hill had been left unprotected through a series of mistakes—wrong assumptions, the failure to communicate clearly, and the failure to check. The situation was critical.

Realizing the danger, COL Vincent ordered his brigade to occupy Little Round Top. He positioned the 20th Maine, commanded by COL Joshua L. Chamberlain, on his brigade's left flank, the extreme left of the Union line. COL Vincent told COL Chamberlain to "hold at all hazards."

On Little Round Top, COL Chamberlain issued his intent and purpose for the mission to the assembled company commanders. He ordered the right flank company to tie in with the 83d Pennsylvania and the left flank company to anchor on a large boulder because the 20th Maine was literally at the end of the line.

COL Chamberlain then showed a skill common to good tactical leaders. He mentally rehearsed possible countermoves against imagined threats to his unit's flank. Since he considered his left flank highly vulnerable, COL Chamberlain sent B Company, commanded by CPT Walter G. Morrill to guard it and "act as the necessities of battle required." The captain positioned his men behind a stone wall, facing the flank of any possible Confederate advance. Fourteen Soldiers from the 2d U.S. Sharpshooters, previously separated from their own unit, joined them.

The 20th Maine had only been in position a few minutes when the Soldiers of the 15th and 47th Alabama attacked. The Confederates, having marched all night, were tired and thirsty, but they attacked ferociously.

The Maine men held their ground until one of COL Chamberlain's officers reported seeing a large body of Confederate Soldiers moving laterally behind the attacking force. COL Chamberlain climbed on a rock and identified a Confederate unit moving around his exposed left flank. He knew that if they outflanked him, his unit would be pushed off its position, facing sure destruction.

COL Chamberlain had to think fast. The tactical manuals he had so diligently studied only offered maneuver solutions, unsuitable for the occupied terrain. He had to create a new stock solution—one that his Soldiers could execute now and under pressure.

Since the 20th Maine was in a defensive line, two ranks deep, and it was threatened by an attack around its left flank, the colonel ordered his company commanders to stretch the line to the left. While keeping up a steady rate of fire, his line ultimately connected with the large boulder he had pointed out earlier. The sidestep maneuver was tricky, but it was a combination of other battle drills his Soldiers knew.

In spite of the terrible noise that confused voice commands, blinding smoke, the cries of the wounded, and the continuing Confederate attack—the Maine men succeeded.

Although COL Chamberlain's thin line was only one rank deep, it now covered twice their normal frontage and was able to throw back the Confederate infantry, assaulting a flank they thought was unprotected.

Despite desperate confederate attempts to break through, the Maine men rallied and held repeatedly. After five desperate encounters, the Maine men were down to one or two rounds per man, and determined Confederates were regrouping for another try.

COL Chamberlain recognized that he could not stay where he was but could not withdraw, either. He decided to attack. His men would have the advantage of attacking down the steep hill, he reasoned, and the Confederates would not expect it. Clearly, he was risking his entire unit, but the fate of the Union Army depended on his men.

The decision left COL Chamberlain with another problem: there was nothing in the tactics book about how to get his unit from current disposition into a firm line of advance. Under tremendous fire in the midst of the battle, COL Chamberlain assembled his commanders. He explained that the regiment's left wing would swing around "like a barn door on a hinge" until it was even with the right wing. Then the entire regiment, bayonets fixed, would charge downhill, staying anchored to the 83d Pennsylvania on the right. The explanation was as simple as the situation was desperate.

When COL Chamberlain gave the order, LT Holman Melcher of F Company leaped forward and led the left wing downhill toward the surprised Confederates. COL Chamberlain had positioned himself at the boulder at the center of the unfolding attack. When his unit's left wing came abreast of the right wing, he jumped off the rock and led the right wing down the hill. The entire regiment was now charging on line, swinging like a great barn door—just as he had intended.

The Alabama Soldiers, stunned at the sight of the charging Union troops, fell back on the positions behind them. There, the 20th Maine's charge might have failed. Just then, CPT Morrill's B Company and the sharpshooters opened fire on the Confederate flank and rear, just as envisioned by COL Chamberlain. The exhausted and shattered Alabama regiments now thought they were surrounded. They broke and ran, not realizing that one more attack would have carried the hill for them.

At the end of the battle, the slopes of Little Round Top were littered with bodies. Saplings halfway up the hill had been sawed in half by weapons fire. A third of the 20th Maine had fallen—130 men out of 386. Nonetheless, the farmers, woodsmen, and fishermen from Maine—under the command of a brave and creative leader, who had anticipated enemy actions, improvised under fire, and applied disciplined initiative in the heat of battle—had fought through to victory.

2-18. Colonel Chamberlain made sure that every man knew what was at stake when his unit prepared for battle at Gettysburg. Prior to the battle, he painstakingly developed his leaders and built his unit into a team with mutual trust between leaders and the subordinates. While teaching and training his Soldiers, he showed respect and compassion for his men and their diverse backgrounds, thus deepening the bond between the commander and his unit. During the battle, he effectively communicated his intent and led by example, with courage and determination. His tactical abilities, intellect, and initiative helped him seize the opportunity and transition from defensive to offensive maneuver, achieving victory over his Confederate opponents. For his actions on 2 July 1863, Colonel Chamberlain received the Medal of Honor.

EXCELLING AT THE CORE LEADER COMPETENCIES

2-19. Leader competence develops from a balanced combination of institutional schooling, self-development, realistic training, and professional experience. Building competence follows a systematic and gradual approach, from mastering individual competencies, to applying them in concert and tailoring them to the situation at hand. Leading people by giving them a complex task helps them develop the confidence and will to take on progressively more difficult challenges.

2-20. Why competencies? Competencies provide a clear and consistent way of conveying expectations for Army leaders. Current and future leaders want to know what to do to succeed in their leadership responsibilities. The core leader competencies apply across all levels of the organization, across leader positions, and throughout careers. Competencies are demonstrated through behaviors that can be readily observed and assessed by a spectrum of leaders and followers: superiors, subordinates, peers, and mentors. This makes them a good basis for leader development and focused multisource assessment and feedback. Figure 2-3 identifies the core leader competencies and their subsets.

	Leads Others	Extends Influence Beyond the Chain of Command	Leads by Example	Communicates
Leads	• Provide purpose, motivation, inspiration. • Enforce standards. • Balance mission and welfare of Soldiers.	• Build trust outside lines of authority. • Understand sphere, means, and limits of influence. • Negotiate, build consensus, resolve conflict.	• Display character. • Lead with confidence in adverse conditions. • Demonstrate competence.	• Listen actively. • State goals for action. • Ensure shared understanding.
	Creates a Positive Environment	Prepares Self	Develops Leaders	
Develops	• Set the conditions for positive climate. • Build teamwork and cohesion. • Encourage initiative. • Demonstrate care for people.	• Be prepared for expected and unexpected challenges. • Expand knowledge. • Maintain self-awareness.	• Assess developmental needs. Develop on the job. • Support professional and personal growth. • Help people learn. • Counsel, coach, and mentor. • Build team skills and processes.	
	Gets Results			
Achieves	• Provide direction, guidance, and priorities. • Develop and execute plans. • Accomplish tasks consistently.			

Figure 2-3. Eight core leader competencies and supporting behaviors

2-21. Leader competencies improve over extended periods. Leaders acquire the basic competencies at the direct leadership level. As the leader moves to organizational and strategic level positions, the competencies provide the basis for leading through change. Leaders continuously refine and extend the ability to perform these competencies proficiently and learn to apply them to increasingly complex situations.

2-22. These competencies are developed, sustained, and improved by performing one's assigned tasks and missions. Leaders do not wait until combat deployments to develop their leader competencies. They use every peacetime training opportunity to assess and improve their ability to lead Soldiers. Civilian leaders also use every opportunity to improve.

2-23. To improve their proficiency, Army leaders can take advantage of chances to learn and gain experience in the leader competencies. They should look for new learning opportunities, ask questions, seek training opportunities, and request performance critiques. This lifelong approach to learning ensures leaders remain viable as a professional corps.

Chapter 3

Leadership Roles, Leadership Levels, and Leadership Teams

3-1. Army leaders of character lead by personal example and consistently act as good role models through a dedicated lifelong effort to learn and develop. They achieve excellence for their organizations when followers are disciplined to do their duty, committed to the Army Values, and feel empowered to accomplish any mission, while simultaneously improving their organizations with focus towards the future.

3-2. As their careers unfold, Army leaders realize that excellence emerges in many shapes and forms. The Army cannot accomplish its mission unless all Army leaders, Soldiers, and civilians accomplish theirs—whether that means filling out a status report, repairing a vehicle, planning a budget, packing a parachute, maintaining pay records, or walking guard duty. The Army consists of more than a single outstanding general or a handful of combat heroes. It relies on hundreds of thousands of dedicated Soldiers and civilians—workers and leaders—to accomplish missions worldwide.

3-3. Each of their roles and responsibilities is unique, yet there are common ways in which the roles of various types of leaders interact. Every leader in the Army is a member of a team, a subordinate, and at some point, a leader of leaders.

ROLES AND RELATIONSHIPS

3-4. When the Army speaks of Soldiers, it refers to commissioned officers, warrant officers, noncommissioned officers (NCOs), and enlisted Soldiers. The term commissioned officer refers to officers serving under a presidential commission in the rank of chief warrant officer 2 through general. An exception is those in the rank of warrant officer 1 (WO1) who serve under a warrant issued by the Secretary of the Army. Army civilians are employees of the Department of the Army and, like all Soldiers, are members of the executive branch of the federal government. All Army leaders, Soldiers, and Army civilians share the same goals: to support and defend the Constitution against all enemies, foreign and domestic, by providing effective Army landpower to combatant commanders and to accomplish their organization's mission in peace and war.

3-5. Although the Army consists of different categories of personnel serving and empowered by different laws and regulations, the roles and responsibilities of Army leaders from all organizations overlap and complement each other. Formal Army leaders come from three different categories: commissioned and warrant officers, noncommissioned officers, and Army civilians.

3-6. Members of all these categories of service have distinct roles in the Army, although duties may sometimes overlap. Collectively, these groups work toward a common goal and should follow a shared institutional value system. Army leaders often find themselves in charge of units or organizations populated with members of all these groups.

COMMISSIONED AND WARRANT OFFICERS

3-7. Commissioned Army officers hold their grade and office under a commission issued under the authority of the President of the United States. The commission is granted on the basis of special trust and confidence placed in the officer's patriotism, valor, fidelity, and abilities. The officer's commission is the grant of presidential authority to direct subordinates and subsequently, an obligation to obey superiors. In the Army, commissioned officers are those who have been appointed to the rank of second lieutenant or higher or promoted to the rank of chief warrant officer 2 or higher.

3-8. Commissioned officers are essential to the Army's organization to command units, establish policy, and manage resources while balancing risks and caring for their people. They integrate collective, leader and Soldier training to accomplish the Army's missions. They serve at all levels, focusing on unit operations and outcomes, to leading change at the strategic levels. Commissioned officers fill command positions. Command makes officers responsible and accountable for everything their command does or fails to do. Command, a legal status held by appointment and grade, extends through a hierarchical rank structure with sufficient authority assigned or delegated at each level to accomplish the required duties.

3-9. Serving as a commissioned officer differs from other forms of Army leadership by the quality and breadth of expert knowledge required, in the measure of responsibility attached, and in the magnitude of the consequences of inaction or ineffectiveness. An enlisted leader swears an oath of obedience to lawful orders, while the commissioned officer promises to, "well and faithfully discharge the duties of the office." This distinction establishes a different expectation for discretionary initiative. Officers should be driven to maintain the momentum of operations, possess courage to deviate from standing orders within the commander's intent when required, and be willing to accept the responsibility and accountability for doing so. While officers depend on the counsel, technical skill, maturity, and experience of subordinates to translate their orders into action, the ultimate responsibility for mission success or failure resides with the commissioned officer in charge.

3-10. The cohorts differ in the magnitude of responsibility vested in them. The life and death decisions conveyed by noncommissioned officers and executed by Soldiers begin with officers. There are different legal penalties assigned for offenses against the authority of commissioned and noncommissioned officers, and there are specific offenses that only an officer can commit. Officers are strictly accountable for their actions. Senior officers bear a particular responsibility for the consequences of their decisions and for the quality of advice given—or not given—to their civilian superiors.

3-11. As they do with all Army leaders, the Army Values guide officers in their daily actions. These values manifest themselves as principles of action. Another essential part of officership is a shared professional identity. This self-concept, consisting of four interrelated identities, inspires and shapes the officer's behavior. These identities are warrior, servant of the Nation, member of a profession, and leader of character. As a warrior and leader of warriors, the officer adheres to the Soldier's Creed and the Warrior Ethos. An officer's responsibility as a public servant is first to the Nation, then to the Army, and then to his unit and his Soldiers. As a professional, the officer is obligated to be competent and stay abreast of changing requirements. As a leader of character, officers are expected to live up to institutional and National ethical values.

3-12. Warrant officers possess a high degree of specialization in a particular field in contrast to the more general assignment pattern of other commissioned officers. Warrant officers command aircraft, maritime vessels, special units, and task organized operational elements. In a wide variety of units and headquarters specialties, warrants provide quality advice, counsel, and solutions to support their unit or organization. They operate, maintain, administer, and manage the Army's equipment, support activities, and technical systems. Warrant officers are competent and confident warriors, innovative integrators of emerging technologies, dynamic teachers, and developers of specialized teams of Soldiers. Their extensive professional experience and technical knowledge qualifies warrant officers as invaluable role models and mentors for junior officers and NCOs.

3-13. Warrant officers fill various positions at company and higher levels. Junior warrants, like junior officers, work with Soldiers and NCOs. While warrant positions are usually functionally oriented, the leadership roles of warrants are the same as other leaders and staff officers. They lead and direct Soldiers and make the organization, analysis, and presentation of information manageable for the commander. Senior warrants provide the commander with the benefit of years of tactical and technical experience.

3-14. As warrant officers begin to function at the higher levels, they become "systems-of-systems" experts, rather than specific equipment experts. As such, they must have a firm grasp of the joint and multinational environments and know how to integrate systems they manage into complex operating environments.

NONCOMMISSIONED OFFICERS

3-15. NCOs conduct the daily operations of the Army. The NCO corps has adopted a vision that defines their role within the Army organization. (See figure 3-1.)

An NCO corps, grounded in heritage, values, and tradition, that embodies the Warrior Ethos; values perpetual learning; and is capable of leading, training, and motivating Soldiers.

We must always be an NCO corps that —
Leads by example.
Trains from experience.
Maintains and enforces standards.
Takes care of Soldiers.
Adapts to a changing world.

Figure 3-1. The NCO vision

3-16. The Army relies on NCOs who are capable of executing complex tactical operations, making intent-driven decisions, and who can operate in joint, interagency, and multinational scenarios. They must take the information provided by their leaders and pass it on to their subordinates. Soldiers look to their NCOs for solutions, guidance, and inspiration. Soldiers can relate to NCOs since NCOs are promoted from the junior enlisted ranks. They expect them to be the buffer, filtering information from the commissioned officers and providing them with the day-to-day guidance to get the job done. To answer the challenges of the contemporary operating environment, NCOs must train their Soldiers to cope, prepare, and perform no matter what the situation. In short, the Army NCO of today is a warrior-leader of strong character, comfortable in every role outlined in the NCO Corps' vision.

3-17. NCO leaders are responsible for setting and maintaining high-quality standards and discipline. They are the standard-bearers. Throughout history, flags have served as rallying points for Soldiers, and because of their symbolic importance, NCOs are entrusted with maintaining them. In a similar sense, NCOs are also accountable for caring for Soldiers and setting the example for them.

3-18. NCOs live and work every day with Soldiers. The first people that new recruits encounter when joining the Army are NCOs. NCOs process Soldiers for enlistment, teach basic Soldier skills, and demonstrate how to respect superior officers. Even after transition from civilian to Soldier is complete, the NCO is the key direct leader and trainer for individual, team, and crew skills at the unit level.

3-19. While preparing Soldiers for the mission ahead, the NCO trainer always stresses the basics of fieldcraft and physical hardening. He knows that the tools provided by technology will not reduce the need for mentally and physically fit Soldiers. Soldiers will continue to carry heavy loads, convoy for hours or days, and clear terrorists from caves and urban strongholds. With sleep often neglected in fast-paced operations, tactical success and failure is a direct correlation to the Soldiers' level of physical fitness. Taking care of Soldiers means making sure they are prepared for whatever challenge lies ahead.

3-20. NCOs have other roles as trainers, mentors, communicators, and advisors. When junior officers first serve in the Army, their NCO helps to train and mold them. When lieutenants make mistakes, seasoned NCOs can step in and guide the young officers back on track. Doing so ensures mission accomplishment and Soldier safety while forming professional and personal bonds with the officers based on mutual trust and common goals. "Watching each other's back" is a fundamental step in team building and cohesion.

3-21. For battalion commanders, the command sergeant major is an important source of knowledge and discipline for all enlisted matters within the battalion. At the highest level, the Sergeant Major of the Army is the Army Chief of Staff's personal advisor, recommending policy to support Soldiers and constantly meeting with and checking Soldiers throughout the Army.

ARMY CIVILIAN LEADERS

3-22. The Army civilian corps consists of experienced personnel committed to serving the Nation. Army civilians are an integral part of the Army team and are members of the executive branch of the federal government. They fill positions in staff and sustaining base operations that would otherwise be filled by military personnel. They provide mission-essential capability, stability, and continuity during war and peace in support of the Soldier. Army civilians take their support mission professionally. Army civilians are committed to selfless service in the performance of their duties as expressed in the Army Civilian Corps Creed. (See figure 3-2.)

I am an Army civilian — a member of the Army team.
I am dedicated to the Army, its Soldiers and civilians.
I will always support the mission.
I provide stability and continuity during war and peace.
I support and defend the Constitution of the United States
and consider it an honor to serve the Nation and its Army.
I live the Army Values of loyalty, duty, respect,
selfless service, honor, integrity, and personal courage.
I am an Army civilian.

Figure 3-2. The Army civilian corps creed

3-23. The major roles and responsibilities of Army civilians include establishing and executing policy; managing Army programs, projects, and systems; and operating activities and facilities for Army equipment, support, research, and technical work. These roles are in support of the organizational Army as well as warfighters based around the world. The main differences between military and civilian leaders are in the provisions of their position, how they obtain their leadership skills, and career development patterns.

3-24. Army civilians' job placement depends on their eligibility to hold the position. Their credentials reflect the expertise with which they enter a position. Proficiency in that position is from education and training they have obtained, prior experiences, and career-long ties to special professional fields. Unlike military personnel, Army civilians do not carry their grade with them regardless of the job they perform. Civilians hold the grade of the position in which they serve. Except for the Commander in Chief (the President of the United States) and Secretary of Defense, civilians do not exercise military command; however, they could be designated to exercise general supervision over an Army installation or activity under the command of a military superior. Army civilians primarily exercise authority based on the position held, not their grade.

3-25. Civilian personnel do not have career managers like their military counterparts, but there are functional proponents for career fields that ensure provisions exist for career growth. Army civilians are free to pursue positions and promotions as they desire. While mobility is not mandatory in all career fields, there are some (and some grade levels) where mobility agreements are required. Personnel policies generally state that civilians should be in positions that do not require military personnel for reasons of law, training, security, discipline, rotation, or combat readiness. While the career civilian workforce brings a wealth of diversity to the Army team, there is also a wealth of knowledge and experience brought to the Army's sustaining base when retired military join the civilian ranks.

3-26. While most civilians historically support military forces at home stations, civilians also deploy with military forces to sustain theater operations. As evidenced by the ever-increasing demands of recent deployments, civilians have served at every level and in every location, providing expertise and support wherever needed. Army civilians support their military counterparts and often remain for long periods within the same organization or installation, providing continuity and stability that the highly mobile personnel management system used for our military rarely allows. However, when the position or mission dictates, Army civilians may be transferred or deployed to meet the needs of the Army.

JOINT AND MULTINATIONAL FORCES

3-27. The Army team may also include embedded joint or multinational forces. Members of these groups, when added to an organization, change both the makeup and the capabilities of the combined team. While leaders may exercise formal authority over joint service members attached to a unit, they must exercise a different form of leadership to influence and guide the behavior of members of allied forces that serve with them. Leaders must adapt to the current operating environment and foster a command climate that includes and respects all members of the Army team.

DEFENSE CONTRACTORS

3-28. A subset of the Army team is contractor personnel. Contractors fill gaps in the available military and Army civilian work force. They also provide services not available through military means to include essential technical expertise to many of our newly fielded weapon systems. Contractor personnel can focus on short-term projects; maintain equipment and aircraft for already over-tasked units; or fill positions as recruiters, instructors, and analysts, freeing up Soldiers to perform Soldier tasks. Contractors used as part of sections, teams, or units must use influence techniques such as those described in Chapter 7 to obtain commitment and compliance as they fulfill their duties or deliver services.

3-29. Managing contractors requires a different leadership approach since they are not part of the military chain-of-command. Contractor personnel should be managed through the terms and conditions set forth in their contract. They do not normally fall under Uniform Code of Military Justice authority. Therefore, it is imperative that Army military and Army civilian leaders ensure that a strong contractor management system is in place in both peacetime and during contingency operations. (See FM 3-100.21 for more information on managing contractor personnel).

SHARED ROLES

3-30. Good leaders wear both Army uniforms and business attire. All leaders take similar oaths upon entry to the Army. These groups work together in a superior-subordinate concept for command positions and formal leadership. Leadership draws on the same aspects of character, using the same competencies regardless of category. The military and civilian functions are complementary and highly integrated. While Soldiers focus on actively fighting and winning in war, the civilian workforce supports all warriors by sustaining operations and helping shape the conditions for mission success. Interdependence and cooperation of these leader categories within the Army make it the multifunctional, highly capable force the Nation depends on.

LEVELS OF LEADERSHIP

NCOs like to make a decision right away and move on to the next thing...so the higher up the flagpole you go, the more you have to learn a very different style of leadership.

Douglas E. Murray
Command Sergeant Major, U.S. Army Reserve (1989)

Figure 3-3. Army leadership levels

3-31. Figure 3-3 shows the three levels of Army leadership: direct, organizational, and strategic. Factors determining a position's leadership level can include the position's span of control, its headquarters level, and the extent of influence the leader holding the position exerts. Other factors include the size of the unit or organization, the type of operations it conducts, the number of people assigned, and its planning horizon.

3-32. Most NCOs, company and field grade officers, and Army civilian leaders serve at the direct leadership level. Some senior NCOs, field grade officers, and higher-grade Army civilians serve at the organizational leadership level. Primarily general officers and equivalent senior executive service Army civilians serve at the organizational or strategic leadership levels.

3-33. Often, the rank or grade of the leader holding a position does not indicate the position's leadership level. That is why Figure 3-3 does not show rank. A sergeant first class serving as a platoon sergeant works at the direct leadership level. If the same NCO holds a headquarters job dealing with issues and policy affecting a brigade-sized or larger organization, that NCO works at the organizational leadership level. However, if the sergeant's primary duty is running a staff section that supports the leaders who run the organization, the NCO is a direct leader.

3-34. It is important to realize that the headquarters echelon alone does not determine a position's leadership level. Leaders of all ranks and grades serve in strategic-level headquarters, but they are not all strategic-level leaders. The responsibilities of a duty position together with the factors listed in paragraph 3-32 usually determine its leadership level. For example, an Army civilian at a post range control facility with a dozen subordinates works at the direct leadership level. An Army civilian deputy garrison commander with a span of influence over several thousand people is an organizational-level leader.

DIRECT LEADERSHIP

3-35. Direct leadership is face-to-face or first-line leadership. It generally occurs in organizations where subordinates are accustomed to seeing their leaders all the time: teams and squads; sections and platoons; companies, batteries, troops, battalions, and squadrons. The direct leader's span of influence may range from a handful to several hundred people. NCOs are in direct leadership positions more often than their officer and civilian counterparts.

3-36. Direct leaders develop their subordinates one-on-one and influence the organization indirectly through their subordinates. For instance, a squadron commander is close enough to the Soldiers to exert direct influence when he visits training or interacts with subordinates during other scheduled functions.

3-37. Direct leaders generally experience more certainty and less complexity than organizational and strategic leaders. Mainly, they are close enough to the action to determine or address problems. Examples of direct leadership tasks are monitoring and coordinating team efforts, providing clear and concise mission intent, and setting expectations for performance.

ORGANIZATIONAL LEADERSHIP

3-38. Organizational leaders influence several hundred to several thousand people. They do this indirectly, generally through more levels of subordinates than do direct leaders. The additional levels of subordinates can make it more difficult for them to see and judge immediate results. Organizational leaders have staffs to help them lead their people and manage their organizations' resources. They establish policies and the organizational climate that support their subordinate leaders.

3-39. Organizational leaders generally include military leaders at the brigade through corps levels, military and civilian leaders at directorate through installation levels, and civilians at the assistant through undersecretary of the Army levels. Their planning and mission focus generally ranges from two to ten years. Some examples of organizational leadership are setting policy, managing multiple priorities and resources, or establishing a long-term vision and empowering others to perform the mission.

3-40. While the same core leader competencies apply to all levels of leadership, organizational leaders usually deal with more complexity, more people, greater uncertainty, and a greater number of unintended consequences. Organizational leaders influence people through policymaking and systems integration rather than through face-to-face contact.

3-41. Getting out of the office and visiting remote parts of their organizations is important for organizational leaders. They make time to get to the field and to the depot warehouses to verify if their staff's reports, e-mails, and briefings match the actual production, the conditions their people face, and their own perceptions of the organization's progress toward mission accomplishment. Organizational leaders use personal observation and visits by designated staff members to assess how well subordinates understand the commander's intent and to determine if there is a need to reinforce or reassess the organization's priorities.

STRATEGIC LEADERSHIP

3-42. Strategic leaders include military and Army civilian leaders at the major command through Department of Defense (DOD) levels. The Army has roughly 600 authorized military and civilian positions classified as senior strategic leaders. Strategic leaders are responsible for large organizations and influence several thousand to hundreds of thousands of people. They establish force structure, allocate resources, communicate strategic vision, and prepare their commands and the Army as a whole for their future roles.

3-43. Strategic leaders work in uncertain environments that present highly complex problems affecting or affected by events and organizations outside the Army. The actions of a geographic combatant commander often have critical impacts on global politics. Combatant commanders command very large, joint organizations with broad, continuing missions. (JP 0-2 and JP 3-0 discuss combatant commands.) There are two different types of combatant commanders:

- Geographic combatant commanders are responsible for a geographic area (called an area of responsibility). For example, the commander of U.S. Central Command is responsible for most of southwestern Asia and part of eastern Africa.
- Functional combatant commanders' responsibilities are not bounded by geography. For example, the commander of the U.S. Transportation Command is responsible for providing integrated land, sea, and air transportation to all Services.

3-44. Strategic leaders apply all core leader competencies they acquired as direct and organizational leaders, while further adapting them to the more complex realities of their strategic environment. Since that environment includes the functions of all Army components, strategic leader decisions must also take into account such things as congressional hearings, Army budgetary constraints, new systems acquisition, civilian programs, research, development, and inter-service cooperation.

3-45. Strategic leaders, like direct and organizational leaders, process information quickly, assess alternatives based on incomplete data, make decisions, and generate support. However, strategic leaders' decisions affect more people, commit more resources, and have wider-ranging consequences in space, time, and political impact, than do decisions of organizational and direct leaders.

3-46. Strategic leaders are important catalysts for change and transformation. Because these leaders generally follow a long-term approach to planning, preparing, and executing, they often do not see their ideas come to fruition during their limited tenure in position. The Army's transformation to more flexible, more rapidly deployable, and more lethal unit configurations, such as brigade combat teams, is a good example of long-range strategic planning. It is a complex undertaking that will require continuous adjustments to shifting political, budgetary, and technical realities. As the transformation progresses, the Army must remain capable of fulfilling its obligation to operate within the full spectrum of military operations on extremely short notice. While the Army relies on many leadership teams, it depends predominantly on organizational leaders to endorse the long-term strategic vision actively to reach all of the Army's organizations.

3-47. Comparatively speaking, strategic leaders have very few opportunities to visit the lowest-level organizations of their commands. That is why they need a good sense of when and where to visit. Because they exert influence primarily through staffs and trusted subordinates, strategic leaders must develop strong skills in selecting and developing talented and capable leaders for critical duty positions.

LEADER TEAMS

3-48. Leaders at all levels recognize the Army is a team as well as a team of teams. These teams interact as numerous functional units, designed to perform necessary tasks and missions that in unison produce the collective effort of all Army components. Everyone belongs to a team, serving as either leader or responsible subordinate. For these teams to function at their best, leaders and followers must develop mutual trust and respect, recognize existing talents, and willingly contribute talents and abilities for the common good of the organization. Leadership within the teams that make up Army usually comes in two forms:

- Legitimate (formal).
- Influential (informal).

FORMAL LEADERSHIP

3-49. Legitimate or formal leadership is granted to individuals by virtue of assignment to positions of responsibility and is a function of rank and experience. The positions themselves are based on the leader's level of job experience and training. One selection process used for the assignment of legitimate authority is the command selection board. Similar to a promotion board, the selection board uses past performance and potential for success to select officers for command positions. NCOs assume legitimate authority when assigned as a platoon sergeant, first sergeant, or command sergeant major. These positions bring with them the duty to recommend disciplinary actions and advancement or promotion.

3-50. The Uniform Code of Military Justice supports military leaders in positions of legitimate authority. Regardless of the quality of leadership exhibited by organizationally appointed leaders, they possess the legal right to impose their will on subordinates, using legal orders and directives.

INFORMAL LEADERSHIP

3-51. Informal leadership can be found throughout organizations, and while it can play an important role in mission accomplishment, it should never undermine legitimate authority. All members of the Army could find themselves in a position to serve as a leader at any time. Informal leadership is not based on any particular rank or position in the organizational hierarchy. It can arise from the knowledge gained from experience and sometimes requires initiative on the part of the individual to assume responsibility not designated to his position. Therefore, even the most junior member may be able to influence the decision of the highest organizational authority. As the final decision maker, the formal leader is ultimately responsible for legitimizing an informal leader's course of action.

IMPLICATIONS FOR ORGANIZATIONAL LEADERS AND UNIT COMMANDERS

3-52. To be effective team builders, organizational leaders and commanders must be able to identify and interact with both formal and informal teams, including—

- The traditional chain of command.
- Chains of coordination directing joint, interagency, and multinational organizations.
- Chains of functional support combining commanders and staff officers.

3-53. Although leading through other leaders is a decentralized process, it does not imply a commander or supervisor cannot step in and temporarily take active control if the need arises. However, bypassing the habitual chain of command should be by exception and focused on solving an urgent problem or guiding an organization back on track with the leader's original guidance.

TEAM STRUCTURES

3-54. There are two leader team categories: horizontal and vertical. Horizontal leader teams can also be either formal (headquarters staffs, major commands) or informal (task forces, advisory boards). Vertical leader teams can be both formal (commanders and subordinates) and informal (members of a career field or functional area). Vertical leader teams often share a common background and function, such as intelligence analysis or logistical support. Vertical and horizontal teams provide structure to organize team training.

3-55. Informal networks often arise both inside and outside formal organizations. Examples of informal networks include people who share experiences with former coworkers or senior NCOs on an installation who collaborate to solve a problem. Although leaders occupy positions of legitimate authority, teams are formed to share information and lessons gained from experience. When groups like this form, they often take on the same characteristics as formally designed organizations. As such, they develop norms unique to their network membership and seek legitimacy through their actions.

3-56. Within the informal network, norms develop for acceptable and unacceptable influence. Studies have shown that groups who do not develop norms of behavior lose their ties and group status.

3-57. The shared leadership process occurs when multiple leaders contribute combined knowledge and individual authority to lead an organization toward a common goal or mission. Shared leadership involves sharing authority and responsibility for decision making, planning, and executing.

3-58. Shared leadership is occurring more frequently at both organizational and strategic levels where leaders of different ranks and positions come together to address specific challenges or missions where pre-established organizational lines of authority may not exist. One such example occurred before Operation Iraqi Freedom when members of multiple components and Services had to work together to support the logistics challenges that lay ahead.

> **Shared Leadership Solves Logistics Challenges**
>
> In the summer of 2002, V Corps hosted a logistics synchronization conference in Germany in anticipation of an impending war.
>
> Representatives from the coalition forces land component, 377th Theater Support Command, and attached units met with the V Corps logistics planners to iron out the details required to move, equip, receive, maintain, sustain, and provide transportation for forces flowing through Kuwait and other locations for the war against Iraq.
>
> Each organization presented its plans and reached a consensus about which component, Service, or provider could best handle each portion of the task.
>
> The mission ahead meant thinking creatively and taking on responsibilities not normally assigned. U.S. Army Central Command–Kuwait (ARCENT-KU) base operations personnel at Camp Doha found their jobs expanding, as they had to collaborate with Kuwaiti bus and trucking companies to provide transportation from the port and airport for the thousands of Soldiers and other Service personnel and contractors that would flow through the country.
>
> The Army and Navy put aside parochial issues to develop a plan to run port operations at the Kuwait Naval Base that would move equipment and personnel smoothly and safely. The Air Force and Army personnel processing units worked in tandem with contractors to design a reception process at Kuwait International Airport.
>
> This was a time when the skill of exercising shared leadership was crucial and unavoidable, given the situation and asset constraints.

3-59. In this example, there were many advantages to using this form of leadership. Alone, each of the organizations might have planned and executed in a vacuum. Together, the group was empowered, calling on their combined base of knowledge and individual subject matter experts to wargame the plan and come up with the best possible courses of action. The result, by the time Operation Iraqi Freedom began, was a cohesive horizontal leader team executing their portions of the plan.

Serving as Responsible Subordinates

3-60. Most leaders are also subordinates within the context of organizations or the institution called the Army. All members of the Army are part of a larger team. A technical supervisor leading several civilian specialists is not just the leader of that group. That team chief also works for someone else and that team has a place in a larger organization.

3-61. Part of being a responsible subordinate implies supporting the chain of command and making sure that the team supports the larger organization and its purpose. Just consider a leader whose team is responsible for handling the pay administration of a large organization. The team chief knows that when the team makes a mistake or falls behind in its work, hard-working Soldiers and civilians pay the price in terms of delayed pay actions. When the team chief introduces a new computer system for handling payroll changes, there is an obligation to try making it succeed, even if the chief initially has doubts that it will work as well as the old one. The team does not exist in a vacuum; it is part of a larger organization, serving many Soldiers, Army civilians, and their families.

3-62. Should the team chief strongly disagree with a superior's implementation concept as project failure that could negatively affect the team's mission and the welfare of many, the chief has an obligation to speak up. The team chief must show the moral courage to voice an opinion in a constructive manner. Disagreement does not imply undermining the chain of command or showing disrespect. Disagreement can lead to a better solution, providing the team chief maintains a positive attitude and offers workable alternatives.

3-63. Ultimately, the discussion must conclude and the team chief should accept a superior's final decision. From that point on, the team chief must support that decision and execute it to the highest of standards. Just imagine what chaos would engulf an organization if subordinates chose freely which orders to obey and which to ignore. In the end, it is important for all leaders to preserve trust and confidence in the chain of command and the collective abilities of the organization.

Leadership without Authority

3-64. Often leadership arises from responsible subordinates who take charge and get the task completed in the absence of clear guidance from superiors. These circumstances arise when situations change or new situations develop for which the leader has not provided guidance or any standing orders for action and cannot be contacted promptly.

3-65. Leadership without authority can originate from one's expertise in a technical area. If others, including those of higher rank, consistently seek a Soldier's or civilian's expertise, that person has an implied responsibility to determine when it is appropriate to take the initiative related to that subject. When leading without designated authority, leaders need to appreciate the potential impact and act to contribute to the team's success. (Appropriate actions are discussed further under the competency of Extends Influence beyond the Chain of Command in Chapter 7.)

3-66. Often leadership without authority arises when one must take the initiative to alert superiors of a potential problem or predict consequences if the organization remains on its current course. Informal leaders without formal authority need to exhibit a leader's image, that of self-confidence and humility.

3-67. Leadership is expected from everyone in the Army regardless of designated authority or recognized position of responsibility. Every leader has the potential to assume ultimate responsibility.

Empowering Subordinates

3-68. Competent leaders know the best way to create a solid organization is to empower subordinates. Give them a task, delegate the necessary authority, and let them do the work. Empowering the team does not mean omitting checks and making corrections when necessary. When mistakes happen, leaders ensure subordinates sort out what happened and why. A quality AAR will help them learn from their mistakes in a positive manner. All Soldiers and leaders err. Good Soldiers and conscientious leaders learn from mistakes.

3-69. Because subordinates learn best by doing, leaders should be willing to take calculated risks and accept the possibility that less experienced subordinates will make mistakes. If subordinate leaders are to grow and develop trust, it is best to let them learn through experience. Good leaders allow space so subordinates can experiment within the bounds of intent-based orders and plans.

3-70. On the opposite end of the spectrum, weak leaders who have not trained their subordinates sometimes insist, "They can't do it without me." Leaders, used to being the center of the attention, often feel indispensable, their battle cry being, "I can't take a day off. I have to be here all the time. I must watch my subordinates' every move, or who knows what will happen?" The fact is that no Army leader is irreplaceable. The Army will not stop functioning just because one leader, no matter how senior or central, steps aside. In combat, the loss of a leader can be a shock to a unit, but the unit must, and will, continue its mission.

Stepping Up to Lead

In the early days of Operation Anaconda, members of the 10th Mountain Division were sent into the Shah-e-kot Valley in eastern Afghanistan. Their mission was to seal off and destroy pockets of Al-Qaeda and Taliban forces. Members of the Afghan National Army assisted by U.S. Special Forces would attack from the north.

CPT Nelson Kraft and his Soldiers from Charlie Company, 1st Battalion, 87th Infantry Regiment were part of the group that would land in the south and wait for them. As soon as the Chinooks carrying the troops landed, the unit found itself in the midst of 100 or more enemy fighters, heavily armed and dug into the cragged mountainsides.

First Platoon was sent up the ridgeline. From their position above the valley, they could hear the mortars advancing closer with each volley. One round hit close to the platoon leader, Lieutenant Brad Maroyka, and wounded him. He gave the order to move, but the next round hit his platoon sergeant. With both leaders out of commission, Kraft radioed SSG Randal Perez, a supply sergeant turned infantryman, and the senior Soldier left standing and told him to take charge.

Reconnaissance photos and intelligence reports had failed to identify this enemy stronghold, but the men of Charlie Company knew they could not run, so they dug in and continued to fight.

Perez did a quick assessment, finding nine of his 26 men wounded. He knew he needed to get them out of the area where they were pinned down. He and five others laid down heavy fire to allow the rest of the team to move to safer ground.

Even though he too was injured, the company first sergeant watched from his position below to see how Perez was handling the pressure. He was glad that the many hours spent at Fort Drum mentoring Perez and teaching him infantry tactics were paying off.

All during the fight, the newly appointed leader controlled his rates of fire, called in targets and kept his men reassured by going helmet to helmet. He rose to the challenge, doing the job of an officer with years of training.

3-71. The company commander, faced with a difficult situation, was confident that the platoon had the depth and experience to regenerate a back-up chain of command. When he picked SSG Perez, he empowered a talented subordinate leader to demonstrate leadership abilities and initiative as envisioned in the commander's intent. The staff sergeant who took charge in the mountains of Afghanistan rose to the occasion because his platoon sergeant had done his job of coaching, mentoring, and counseling him. He showed the courage to expose himself to enemy fire and quickly re-instilled confidence in his Soldiers. He kept reassuring them that things would be okay and kept them informed of the unfolding tactical situation. The staff sergeant understood the essentials of leading by example under adverse situations. When things were not going exactly in accordance with the plan, the acting platoon leader inspired his unit to persevere and bond against all odds. He mobilized a tremendous psychological force—morale.

3-72. Leaders will have numerous roles and responsibilities throughout their time spent serving. Some will be commanders, staff officers, or senior civilians. Some will serve as platoon sergeants or first sergeants. Others will be recruiters and instructors, leading through example, and seeking out and training tomorrow's leaders. Duty assignments may include time on a joint task force or as the member of a team seeking answers to future challenges. Whatever their role, Army leaders must have the character, presence, and intellect to do whatever is asked of them.

PART TWO

The Army Leader: Person of Character, Presence, and Intellect

Army leadership doctrine concerns itself with all aspects of leadership, the most important of which is the Army leader. Part Two examines that person and highlights critical attributes that all Army leaders can bring to bear, in order to reach their full professional potential on a career path from direct leader to strategic leader. It demonstrates that when Soldiers and Army civilians begin as leaders, they bring certain values and attributes, such as family-ingrained values and the aptitude for certain sports or intellectual abilities, such as learning foreign languages. Army institutional training, combined with education, training, and development on the job, aims at using these existing qualities and potential to develop a well-rounded leader with sets of desired attributes forming the leader's character, presence, and intellect. Development of the desired attributes requires that Army leaders pay attention to them through consistent self-awareness and lifelong learning. Appendix A lists the set of leader attributes and core leader competencies.

Chapter 4

Leader Character

Just as fire tempers iron into fine steel so does adversity temper one's character into firmness, tolerance, and determination....

Margaret Chase Smith
Lieutenant Colonel, U.S. Air Force Reserve and United States Senator
Speech to graduating women naval officers at Newport, RI (1952)

4-1. Character, a person's moral and ethical qualities, helps determine what is right and gives a leader motivation to do what is appropriate, regardless of the circumstances or the consequences. An informed ethical conscience consistent with the Army Values strengthens leaders to make the right choices when faced with tough issues. Since Army leaders seek to do what is right and inspire others to do the same, they must embody these values.

4-2. American Soldier actions during Operation Desert Storm speak about values, attributes, and character.

Soldier Shows Character and Discipline

On the morning of 28 February 1991, about a half-hour prior to the cease-fire, a T-55 tank pulled up in front of a U.S Bradley unit, which immediately prepared to engage with TOW missiles. A vehicle section consisting of the platoon sergeant and his wingman tracked the Iraqi tank, ready to unleash two deadly shots.

> Suddenly, the wingman saw the T-55 tank stop and a head popped up from the commander's cupola. The wingman immediately radioed to his platoon sergeant to hold his fire, believing that the Iraqi was about to dismount the vehicle, possibly to surrender.
>
> The Iraqi tank crew jumped off their vehicle and ran behind a sand dune. Sensing something was not right, the platoon sergeant immediately instructed his wingman to investigate the area around a nearby dune, while he provided cover with his weapons. To everyone's surprise, the wingman and his crew soon discovered 150 enemy combatants ready to surrender.
>
> To deal with this vast number of prisoners, the American unit lined them up and ran them through a gauntlet to disarm them and check them for items of intelligence value. Then the unit called for prisoner of war handlers to pick them up.
>
> Before moving on, the platoon sergeant had to destroy the T-55 tank. Before blowing it in place, the NCO instructed it moved behind a sand berm to protect his people and the prisoners from the shrapnel of the tank's on-board munitions.
>
> When the tank suddenly exploded and the small arms cooked-off, sounding as if small arms were fired, the prisoners panicked, believing the Soldiers would shoot them. Quickly, the Soldiers communicated that this would not happen, one of them telling the Iraqis, "Hey, we're from America, we don't shoot our prisoners!"

4-3. The Soldier's comment captures the essence of Army values-based character. There is a direct connection between the leader's character and his actions. Character, discipline, and good judgment allowed the platoon sergeant and his wingman to hold fire for the proper surrender of enemy combatants. Sound reasoning and ethical considerations guided the platoon sergeant in his decision to safeguard his own men and the prisoners from the dangerous debris caused by the T-55's explosion. He and his Soldiers safeguarded the Army Values and standards of conduct by reassuring the prisoners that they would be unharmed.

4-4. Character is essential to successful leadership. It determines who people are and how they act. It helps determine right from wrong and choose what is right. The factors, internal and central to a leader, which make up the leader's core are—

- Army Values.
- Empathy.
- Warrior Ethos.

ARMY VALUES

4-5. Soldiers and Army civilians enter the Army with personal values developed in childhood and nurtured over many years of personal experience. By taking an oath to serve the Nation and the institution, one also agrees to live and act by a new set of values—Army Values. The Army Values consist of the principles, standards, and qualities considered essential for successful Army leaders. They are fundamental to helping Soldiers and Army civilians make the right decision in any situation.

4-6. The Army Values firmly bind all Army members into a fellowship dedicated to serve the Nation and the Army. They apply to everyone, in every situation, anywhere in the Army. The trust Soldiers and civilians have for each other and the trust of the American people, all depend on how well a Soldier embodies the Army Values.

4-7. The Army recognizes seven values that must be developed in all Army individuals. It is not coincidence that when reading the first letters of the Army Values in sequence they form the acronym "LDRSHIP":

- Loyalty.
- Duty.

- Respect.
- Selfless service.
- Honor.
- Integrity.
- Personal courage.

Loyalty

Bear true faith and allegiance to the U.S. Constitution, the Army, your unit, and other Soldiers.

> *Loyalty is the big thing, the greatest battle asset of all. But no man ever wins the loyalty of troops by preaching loyalty. It is given him by them as he proves his possession of the other virtues.*
>
> Brigadier General S. L. A. Marshall
> *Men Against Fire* (1947)

4-8. All Soldiers and government civilians swear a sacred oath to support and defend the Constitution of the United States. The Constitution established the legal basis for the existence of our Army. Article I, Section 8, outlines congressional responsibilities regarding America's armed forces. As a logical consequence, leaders as members of the armed forces or government civilians have an obligation to be faithful to the Army and its people.

4-9. Few examples better illustrate loyalty to country, the Army, its people, and self better than that of World War II General Jonathan Wainwright.

Loyal in War and in Captivity

The Japanese invaded the Philippines in December 1941. In March 1942, GEN Douglas MacArthur left his Philippine command and evacuated to Australia. Although GEN MacArthur intended to stay in command from Australia, GEN Jonathan Wainwright, a tall, thin and loyal general officer assumed full command from the Malinta Tunnel on Corregidor, while Major General Edward King replaced Wainwright as commander of the American Forces and Filipino Scouts defending Bataan.

Soon, the Japanese grip on the islands tightened and the Philippine defenders at Bataan were surrounded and without any support other than artillery fire from Corregidor. Disease, exhaustion, and malnutrition ultimately accomplished what thousands of Japanese soldiers had not done for 90 days—Bataan was lost.

When Bataan fell to the Japanese, more than 12,000 Filipino Scouts and 17,000 Americans became prisoners. On the initial march to Camp O'Donnell, the Japanese beheaded many who became too weak to continue the trip. Other prisoners were used for bayonet practice or pushed to their deaths from cliffs.

The situation at Corregidor was no better. Soldiers were weary, wounded, malnourished, and diseased. GEN Wainwright directed the defenses with the limited resources available. Wainwright made frequent visits to the front to check on his men and to inspire them personally. He never feared coming under direct fire from enemy soldiers. A tenacious warrior, he was used to seeing men next to him die and had often personally returned fire on the enemy.

GEN Wainwright was a unique kind of frontline commander—a fighting general who earned the loyalty of his troops by sharing their hardships.

GEN Wainwright and his steadfast troops at Corregidor were the last organized resistance on Luzon. After holding the Japanese against impossible odds for a full six months, Wainwright had exhausted all possibilities—no outside help could be expected.

> On 6 May 1942, GEN Wainwright notified his command of his intent to surrender and sent a message to the President of the United States to explain the painful decision. He was proud of his country and his men and he had been forthright and loyal to both. His Soldiers had come to love, admire, and willingly obey the fighting general. President Roosevelt reassured GEN Wainwright of the Nation's loyalty and in one of his last messages to him wrote: "You and your devoted followers have become the living symbol of our war aims and the guarantee of victory."
>
> Following the surrender, the Japanese shipped the defenders of Corregidor across the bay to Manila where they were paraded in disgrace. To humiliate him personally, GEN Wainwright was forced to march through his defeated Soldiers. Despite their wounds, their illness, their broken spirit, and shattered bodies, Wainwright's Soldiers once again demonstrated their loyalty and respect for their leader. As he passed among their ranks, the men struggled to their feet and saluted.
>
> During his more than three years of captivity as the highest-ranking and oldest American prisoner of war in World War II, GEN Wainwright kept faith and loyalty with his fellow prisoners and suffered many deprivations, humiliation, abuse, and torture.
>
> Despite his steadfast posture in captivity, GEN Wainwright feared the moment of his return to America, expecting to be considered a coward and a traitor for his surrender at Corregidor. Americans at home had not forgotten and remained loyal to the fighting general and his courageous troops. To honor him and his men, GEN Wainwright stood behind GEN of the Army MacArthur together with British GEN Percival, during the signing of Japan's official surrender on board the battleship USS Missouri, on 2 September 1945.
>
> GEN Jonathan Wainwright subsequently returned home not to experience shame, but a hero's welcome. During a surprise ceremony on 10 September 1945, President Truman awarded Jonathan Wainwright the Medal of Honor.

4-10. The bond of loyalty not only encompasses the institution and the Nation's legal foundation, but also reaches into every unit and organization. At unit and organizational levels, loyalty is a two-way commitment between leaders and subordinates.

There is a great deal of talk about loyalty from the bottom to the top. Loyalty from the top down is even more necessary and much less prevalent.

General George S. Patton
War As I Knew It (1947)

4-11. The loyalty of subordinates is a gift given when a leader deserves it. Leaders earn subordinates' loyalty by training them well, treating them fairly, and living the Army Values. Leaders who are loyal to their subordinates never let Soldiers be misused or abused. Subordinates who believe in their leaders will stand with them no matter how difficult the situation.

4-12. Research and historical data agree that Soldiers and units fight for each other. Loyalty bonds them together. Without a doubt, the strongest bonds emerge when leading people in combat. While combat is the most powerful bonding experience, good units can build loyalty and trust during peacetime.

4-13. Loyalty and trust are extremely important ingredients for the successful day-to-day operations of all organizations, many of them a mix of Army civilians and Soldiers. The logistical and political demands of modern war have greatly expanded the roles of civilians, regardless if employed by contractors or the Department of the Army. Whether stationed at home or in forward-deployed operational theaters, their contributions are vital to many mission accomplishments. They are loyal partners of the Army team, running logistical convoys, repairing infrastructure, maintaining complex equipment, and feeding Soldiers.

4-14. To create strong organizations and tight-knit small-unit brotherhoods, all team members must embrace loyalty—superiors, subordinates, peers, civilians, and Soldiers. Loyalty encompasses all Army components, including the National Guard and Army Reserve who shoulder an increasingly growing share of the Army's long-term operational commitments. Ultimately, the bonds of loyalty also extend to other

Services. While many think they can easily go it alone, the reality of modern, multidimensional war shows that joint capabilities are essential to successful mission outcomes.

DUTY

Fulfill your obligations.

> *I go anywhere in the world they tell me to go, any time they tell me to, to fight anybody they want me to fight. I move my family anywhere they tell me to move, on a day's notice, and live in whatever quarters they assign me. I work whenever they tell me to work.... And I like it.*
>
> James H. Webb
> Former U.S. Marine and Secretary of the Navy (1987-1988)

4-15. Duty extends beyond everything required by law, regulation, and orders. Professionals work not just to meet the minimum standard, but consistently strive to do their very best. Army leaders commit to excellence in all aspects of their professional responsibility.

4-16. Part of fulfilling duty is to exercise initiative—anticipating what needs to be done before being told what to do. Army leaders exercise initiative when they fulfill the purpose, not merely the letter, of the tasks they have been assigned and the orders they have received. The task is not complete until the intended outcome is achieved. When a platoon sergeant tells a squad leader to inspect weapons, the squad leader only fulfills a minimum obligation when checking weapons. If the squad leader finds weapons that are not clean or serviced, a sense of duty alerts the leader to go beyond the platoon sergeant's instructions. To fulfill that duty, squad leaders must correct the problem and ensure that all the unit's weapons are up to standard. When leaders take initiative, they also take full responsibility for their actions and those of their subordinates. Conscientiousness is a human trait where duty is internalized. Conscientiousness means having a high sense of responsibility for personal contributions to the Army, demonstrated through dedicated effort, organization, thoroughness, reliability, and practicality. Conscientiousness consistently alerts the leader to do what is right—even when tired or demoralized.

4-17. In rare cases, a leader's sense of duty also has to detect and prevent an illegal order. Duty requires refusal to obey it—leaders have no choice but to do what is ethically and legally right.

RESPECT

Treat people as they should be treated.

> *The discipline which makes the soldiers of a free country reliable in battle is not to be gained by harsh or tyrannical treatment. On the contrary, such treatment is far more likely to destroy than to make an army. It is possible to impart instruction and to give commands in such manner and such a tone of voice to inspire in the soldier no feeling but an intense desire to obey, while the opposite manner and tone of voice cannot fail to excite strong resentment and a desire to disobey. The one mode or the other of dealing with subordinates springs from a corresponding spirit in the breast of the commander. He who feels the respect which is due to others cannot fail to inspire in them regard for himself, while he who feels, and hence manifests, disrespect toward others, especially his inferiors, cannot fail to inspire hatred against himself.*
>
> Major General John M. Schofield
> Address to the United States Corps of Cadets, 11 August 1879

4-18. Respect for the individual is the basis for the rule of law—the very essence of what the Nation stands for. In the Army, respect means treating others as they should be treated. This value reiterates that people are the most precious resource and that one is bound to treat others with dignity and respect.

4-19. Over the course of history, America has become more culturally diverse, requiring Army leaders to deal with people from a wider range of ethnic, racial, and religious backgrounds. An Army leader should prevent misunderstandings arising from cultural differences. Actively seeking to learn about people whose culture is different can help to do this. Being sensitive to other cultures will aid in mentoring, coaching, and counseling subordinates. This demonstrates respect when seeking to understand their background, see things from their perspective, and appreciate what is important to them.

4-20. Army leaders should consistently foster a climate in which everyone is treated with dignity and respect, regardless of race, gender, creed, or religious belief. Fostering a balanced and dignified work climate begins with a leader's personal example. How a leader lives the Army Values shows subordinates how they should behave. Teaching values is one of a leader's most important responsibilities. It helps create a common understanding of the Army Values and expected standards.

SELFLESS SERVICE

Put the welfare of the Nation, the Army, and subordinates before your own.

> *...[A]sk not what your country can do for you; ask what you can do for your country.*
>
> John F. Kennedy
> Inaugural speech as 35th President of the United States (1961)

4-21. The military is often referred to as "the Service." Members of the Army serve the United States of America. Selfless service means doing what is right for the Nation, the Army, the organization, and subordinates. While the needs of the Army and the Nation should come first, it does not imply family or self-neglect. To the contrary, such neglect weakens a leader and can cause the Army more harm than good.

4-22. A strong but harnessed ego, high self-esteem, and a healthy ambition can be compatible with selfless service, as long as the leader treats his people fairly and gives them the credit they deserve. The leader knows that the Army cannot function except as a team. For a team to excel, the individual must give up self-interest for the good of the whole.

4-23. Selfless service is not only expected of Soldiers. Civilians, supporting many of the Army's most critical missions, should display the same value. During Operation Desert Storm, many of the civilians deployed to Southwest Asia volunteered to serve there, filling vital roles in supporting Soldiers and operations.

4-24. On 11 September 2001, after the attack on the Pentagon, that selfless team effort between military personnel and civilian workers did not come as a surprise. Civilians and Soldiers struggled side-by-side to save each other's lives, while together they ensured that critical operations around the world continued without loss of command and control.

4-25. Often, the need for selflessness is not limited to combat or emergencies. Individuals continue to place the Army's needs above their own as retirees volunteer for recall, members of the Reserve Components continue to serve beyond their mandatory service dates, and Army civilians volunteer for duty in combat zones.

HONOR

Live up to all the Army Values.

> *War must be carried on systematically, and to do it you must have men of character activated by principles of honor.*
>
> George Washington
> Commander, Continental Army (1775-81) and President of the United States (1789-97)

4-26. Honor provides the moral compass for character and personal conduct for all members of the Army. Honor belongs to those living by words and actions consistent with high ideals. The expression "honorable person" refers to the character traits an individual possesses that the community recognizes and respects.

4-27. Honor is the glue that holds the Army Values together. Honor requires a person to demonstrate continuously an understanding of what is right. It implies taking pride in the community's acknowledgment of that reputation. Military ceremonies recognizing individual and unit achievements demonstrate and reinforce the importance the Army places on honor.

4-28. The Army leader must demonstrate an understanding of what is right and take pride in that reputation by living up to the Army Values. Living honorably, in line with the Army Values, sets an example for every member of the organization and contributes to an organization's positive climate and morale.

4-29. How leaders conduct themselves and meet obligations define them as persons and leaders. In turn, how the Army meets the Nation's commitments defines the Army as an institution. Honor demands putting the Army Values above self-interest and above career and personal comfort. For Soldiers, it requires putting the Army Values above self-preservation. Honor gives the strength of will to live according to the Army Values, especially in the face of personal danger. It is not coincidence that our military's highest award is the Medal of Honor. Its recipients clearly went beyond what is expected and beyond the call of duty.

Honor, Courage, and Selfless Service in Korea

On 14 June 1952 SGT David B. Bleak, a medical aidman in Medical Company, 223rd Infantry Regiment, 40th Infantry Division volunteered to accompany a combat patrol tasked to capture enemy forces for interrogation. While moving up the rugged slope of Hill 499, near Minari-gol, Korea, the patrol came under intense automatic weapons and small arms fire several times, suffering several casualties. An enemy group fired at SGT Bleak from a nearby trench while he tended the wounded.

Determined to protect the wounded, the brave aidman faced the enemy. He entered the trench and killed two enemy soldiers with his bare hands and a third with his trench knife. While exiting, SGT Bleak detected a concussion grenade as it fell in front of a comrade. Bleak quickly shifted to shield the man from the blast.

Disregarding his own injury, he carried the most severely wounded comrade down a hillside. Attacked by two enemy soldiers with bayonets, Bleak lowered the wounded man and put both adversaries out of action by slamming their heads together. He then carried the wounded American Soldier to safety.

SGT Bleak's courageous actions saved fellow Soldiers' lives and preserved the patrol's combat effectiveness. For his actions, President Dwight D. Eisenhower awarded him the Medal of Honor on 27 October 1953.

INTEGRITY

Do what's right—legally and morally.

No nation can safely trust its martial honor to leaders who do not maintain the universal code which distinguishes between those things that are right and those things that are wrong.

General Douglas MacArthur
Patriot Hearts (2000)

4-30. Leaders of integrity consistently act according to clear principles, not just what works now. The Army relies on leaders of integrity who possess high moral standards and who are honest in word and deed. Leaders are honest to others by not presenting themselves or their actions as anything other than what they are, remaining committed to the truth.

4-31. Here is how a leader stands for the truth: if a mission cannot be accomplished, the leader's integrity requires him to inform the chain of command. If the unit's operational readiness rate is truly 70 percent, despite the senior commander's required standard of 90 percent, a leader of integrity will not instruct subordinates to adjust numbers. It is the leader's duty to report the truth and develop solutions to meet the standard with honor and integrity. Identifying the underlying maintenance issues and raising the quality bar could ultimately save Soldiers' lives.

4-32. If leaders inadvertently pass on bad information, they should correct it as soon as they discover the error. Leaders of integrity do the right thing not because it is convenient or because they have no other choice. They choose the path of truth because their character permits nothing less.

4-33. Serving with integrity encompasses more than one component. However, these components are dependant on whether the leader inherently understands right versus wrong. Assuming the leader can make the distinction, a leader should always be able to separate right from wrong in every situation. Just as important, that leader should do what is right, even at personal cost.

4-34. Leaders cannot hide what they do, but must carefully decide how to act. Army leaders are always on display. To instill the Army Values in others, leaders must demonstrate them personally. Personal values may extend beyond the Army Values, to include such things as political, cultural, or religious beliefs. However, as an Army leader and a person of integrity, these values should reinforce, not contradict, the Army Values.

4-35. Conflicts between personal and Army Values should be resolved before a leader becomes a morally complete Army leader. If in doubt, a leader may consult a mentor with respected values and judgment.

PERSONAL COURAGE

Face fear, danger, or adversity (physical and moral).

> *Courage is doing what you're afraid to do. There can be no courage unless you're scared.*
>
> Captain Eddie Rickenbacker
> U.S. Army Air Corps, World War I

4-36. As the Army Air Corps World War I fighter ace, Captain Eddie Rickenbacker, put it—personal courage is not the absence of fear. It is the ability to put fear aside and do what is necessary. Personal courage takes two forms: physical and moral. Good leaders demonstrate both.

4-37. Physical courage requires overcoming fears of bodily harm and doing one's duty. It triggers bravery that allows a Soldier to take risks in combat in spite of the fear of wounds or even death. One lieutenant serving during World War II displayed such courage despite serving in a time when he and his fellow African-American Soldiers were not fully recognized for their actions.

Courage and Inspiration for Soldiers Then and Now

Of all the Medals of Honor awarded during World War II, none went to an African-American. In 1993, the Army contracted Shaw University in Raleigh, North Carolina, to research racial disparities in the selection of Medal of Honor recipients. As a result, the Army ultimately decided to recommend seven for the award.

Fifty-two years after they earned them, the medals were awarded along with the nation's silent apology for being ignored by the once-segregated Army. The only soldier still alive to receive his award was Vernon J. Baker, an exceptionally courageous and inspirational leader.

On 5 and 6 April 1945, Second Lieutenant Baker of the 370th Infantry Regiment had demonstrated leadership by example near Viareggio, Italy, during his company's attack against strongly entrenched German positions in mountainous terrain.

When his company was stopped by fire from several machine gun emplacements, LT Baker crawled to one position and destroyed it, killing three German soldiers. He then attacked an enemy observation post and killed two occupants. With the aid of one of his men, LT Baker continued the advance and destroyed two more machine gun nests, killing or wounding the soldiers occupying these positions. After consolidating his position, LT Baker finally covered the evacuation of the wounded personnel of his unit by occupying an exposed position and drawing the enemy's fire.

On the night following his heroic combat performance, LT Baker again volunteered to lead a battalion advance toward his division's objective through enemy mine fields and heavy fire. Two-thirds of his company was wounded or dead and there were no reinforcements in sight. His commander ordered a withdrawal. Baker, in tears protested, "Captain, we can't withdraw. We must stay here and fight it out."

LT Baker stands as an inspiration to the many African-American Soldiers who served with him and since that time. He stood courageously against the enemy and stood proudly to represent his fallen comrades when he received his Medal of Honor.

Long after he saw combat in Italy, Vernon J. Baker still thought of his black comrades who died around him as they awaited reinforcements that never came. In a CNN interview, he summed up his feelings with the following modest words: "This day will vindicate those men and make things right."

4-38. Moral courage is the willingness to stand firm on values, principles, and convictions. It enables all leaders to stand up for what they believe is right, regardless of the consequences. Leaders, who take full responsibility for their decisions and actions, even when things go wrong, display moral courage.

4-39. General Dwight D. Eisenhower was a leader of great moral courage during his service as the Supreme Commander of Allied Forces Europe. He displayed this moral courage in a handwritten note he prepared for public release, in case the Normandy landings failed.

> *Our landings in the Cherbourg-Havre area have failed to gain a satisfactory foothold and I have withdrawn the troops. My decision to attack at this time and place was based upon the best information available. The troops, the air, and the Navy did all that bravery and devotion to duty could do. If any blame or fault attaches to the attempt it is mine alone—June 5.*

4-40. Moral courage also expresses itself as candor. Candor means being frank, honest, and sincere with others. It requires steering clear of bias, prejudice, or malice even when it is uncomfortable or may seem better to keep quiet.

> *The concept of professional courage does not always mean being as tough as nails, either. It also suggests a willingness to listen to the soldiers' problems, to go to bat for them in a tough situation and it means knowing just how far they can go. It also means being willing to tell the boss when he is wrong.*
>
> William Connelly
> Sergeant Major of the Army (1979-1983)

4-41. One can observe candor when a company commander calmly explains to the first sergeant that a Soldier should receive a lower-level punishment, although the first sergeant insists on a company-grade Article 15. Likewise, a candid first sergeant respectfully points out a company commander might be overreacting for ordering remedial weekend maintenance for the entire company, when only one platoon actually failed inspection. Trust relationships between leaders and subordinates rely on candor. Without it, subordinates will not know if they have met the standard and leaders will not know what is going on in their organization.

EMPATHY

4-42. Army leaders show a propensity to share experiences with the members of their organization. When planning and deciding, try to envision the impact on Soldiers and other subordinates. The ability to see something from another person's point of view, to identify with and enter into another person's feelings and emotions, enables the Army leader to better care for civilians, Soldiers, and their families.

4-43. Competent and empathetic leaders take care of Soldiers by giving them the training, equipment, and all the support they need to keep them alive in combat and accomplish the mission. During wartime and difficult operations, empathetic Army leaders share the hardships with their people to gauge if their plans and decisions are realistic. Competent and empathetic leaders also recognize the need to provide Soldiers and civilians with reasonable comforts and rest periods to maintain good morale and mission effectiveness. When a unit or organization suffers injuries or death, empathetic Army leaders can help ease the trauma and suffering in the organization to restore full readiness as quickly as possible.

4-44. Modern Army leaders recognize that empathy also includes nourishing a close relationship between the Army and Army families. To build a strong and ready force, Army leaders at all levels promote self-sufficient and healthy families. Empathy for families includes allowing Soldiers recovery time from difficult missions, protecting leave periods, permitting critical appointments, as well as supporting events that allow information exchange and family teambuilding.

4-45. The requirement for leader empathy extends beyond civilians, Soldiers, and their families. Within the larger operational environment, leader empathy may be helpful when dealing with local populations and prisoners of war. Providing the local population within an area of operations with the necessities of life often turns an initially hostile disposition into one of cooperation.

THE WARRIOR ETHOS

4-46. General Eric Shinseki, former Army Chief of Staff, described the need for a common Warrior Ethos with emphasis on the uniformed members of the Army team:

> *Every organization has an internal culture and ethos. A true warrior ethos must underpin the Army's enduring traditions and values.... Soldiers imbued with an ethically grounded warrior ethos clearly symbolize the Army's unwavering commitment to the nation we serve. The Army has always embraced this ethos but the demands of Transformation will require a renewed effort to ensure that all Soldiers truly understand and embody this warrior ethos.*

4-47. The Warrior Ethos refers to the professional attitudes and beliefs that characterize the American Soldier. It echoes through the precepts of the Code of Conduct and reflects a Soldier's selfless commitment to the Nation, mission, unit, and fellow Soldiers. The Warrior Ethos was developed and sustained through discipline, commitment to the Army Values, and pride in the Army's heritage. Lived by Soldiers and supported by dedicated Army civilians, a strong Warrior Ethos is the foundation for the winning spirit that permeates the institution.

4-48. U.S Army Soldiers embrace the Warrior Ethos as defined in the Soldier's Creed. (See figure 4-1.)

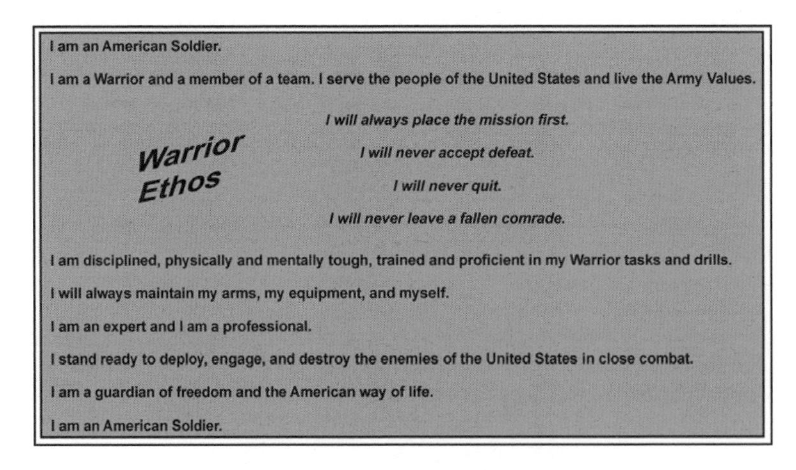

Figure 4-1. The Soldier's Creed

4-49. The Warrior Ethos is more than persevering in war. It fuels the fire to fight through any demanding conditions—no matter the time or effort required. It is one thing to make a snap decision to risk one's life for a brief period. It is quite another to sustain the will to win when the situation looks hopeless and shows no indication of getting better, when being away from home and family is already a profound hardship. The Soldier who jumps on a grenade to save comrades is courageous without question—that action requires great mental and physical courage. Pursuing victory over extended periods with multiple deployments requires this deep moral courage, one that focuses on the mission.

> *Wars may be fought with weapons, but they are won by men. It is the spirit of the men who follow and of the man who leads that gains the victory.*
>
> General George S. Patton
> *Cavalry Journal* (1933)

4-50. The actions of all who have fought courageously in wars past exemplify the essence of the Army's Warrior Ethos. Developed through discipline, commitment to the Army Values, and knowledge of the Army's proud heritage, the Warrior Ethos makes clear that military service is much more than just another job. It is about the warrior's total commitment. It is the Soldiers' absolute faith in themselves and their comrades that makes the Army invariably persuasive in peace and invincible in war. The Warrior Ethos forges victory from the chaos of battle. It fortifies all leaders and their people to overcome fear, hunger, deprivation, and fatigue. The Army wins because it fights hard and with purpose. It fights hard because it trains hard. Tough training is the path to winning at the lowest cost in human sacrifice.

4-51. The Warrior Ethos is a component of character. It shapes and guides what a Soldier does. It is linked tightly to the Army Values such as personal courage, loyalty to comrades, and dedication to duty. During the Korean War, one leader displayed these traits and surpassed traditional bounds of rank to lead his Soldiers.

Task Force Kingston

LT Joseph Kingston, a boyish-looking platoon leader in K Company, 3d Battalion, 32d Infantry, was commanding the lead element for his battalion's move northward. The terrain was mountainous in that part of Korea, the weather bitterly cold—the temperature often below zero—and the cornered enemy still dangerous.

LT Kingston inched his way forward, the battalion gradually adding elements to his force. Soon, he had anti-aircraft jeeps mounted with quad .50 caliber machine guns, a tank, a squad (later a platoon) of engineers, and an artillery forward observer under his control. Some of the new attachments were commanded by lieutenants who outranked him, as did the tactical air controller—a captain. LT Kingston remained in command, and battalion headquarters began referring to his growing force as, "Task Force Kingston."

Bogged down in Yongsong-ni with casualties mounting, Task Force Kingston received reinforcements that brought its strength to nearly 300. 1LT Kingston's battalion commander wanted him to remain in command, even though he pushed forward several more officers who outranked LT Kingston. One of the attached units was a rifle company, commanded by a captain. Nonetheless, the cooperative command arrangement worked—because LT Kingston was a very competent leader.

Despite tough fighting, the force advanced. Hit while leading an assault on one enemy stronghold, Kingston managed to toss a grenade, just as a North Korean soldier fired a shot that glanced off his helmet. The Lieutenant's resilience and personal courage inspired every Soldier from the wide array of units under his control.

Task Force Kingston succeeded in battle because of a competent young leader who inspired his people by demonstrating many attributes common to the Warrior Ethos and the Army Values that the Army currently espouses.

4-52. The Warrior Ethos requires unrelenting and consistent determination to do what is right and to do it with pride across the spectrum of conflicts. Understanding what is right requires respect for both comrades and all people involved in complex missions, such as stability and reconstruction operations. Ambiguous situations, such as when to use lethal or nonlethal force, are a test for the leader's judgment and discipline. The Warrior Ethos helps create a collective commitment to win with honor.

4-53. The Warrior Ethos is crucial but also perishable. Consequently, the Army must continually affirm, develop, and sustain it. The martial ethic connects American warriors of today with those whose sacrifices have sustained our very existence since America's founding. The Army's continuing drive to be the best, to triumph over all adversity, and to remain focused on mission accomplishment, does more than preserve the Army's institutional culture—it sustains the Nation.

4-54. Actions that safeguard and sustain the Nation occur everywhere there are Soldiers and civilian members of the Army team. All that tireless motivation comes in part from the cohesion that springs from the Warrior Ethos. Soldiers fight for each other and their loyalty runs front to rear as well as left to right. Mutual support is a defining characteristic of Army culture, present regardless of time or place.

CHARACTER DEVELOPMENT

4-55. People join the Army as Soldiers and Army civilians with their character, pre-shaped by their background, beliefs, education, and experience. An Army leader's job would be simpler if merely checking the team member's personal values against the Army Values and developing a simple plan to align them sufficed. Reality is much different. Becoming a person of character and a leader of character is a career-long process involving day-to-day experience, education, self-development, developmental counseling, coaching, and mentoring. While individuals are responsible for their own character development, leaders are responsible for encouraging, supporting, and assessing the efforts of their people. Leaders of character can develop only through continual study, reflection, experience, and feedback. Leaders hold themselves and subordinates to the highest standards. The standards and values then spread throughout the team, unit, or organization and ultimately throughout the Army.

4-56. Doing the right thing is good. Doing the right thing for the right reason and with the right goal is better. People of character must possess the desire to act ethically in all situations. One of the Army leader's primary responsibilities is to maintain an ethical climate that supports development of such character. When an organization's ethical climate nurtures ethical behavior, people will, over time, think, feel, and act ethically. They will internalize the aspects of sound character.

CHARACTER AND BELIEFS

4-57. Beliefs matter because they help people understand their experiences. Those experiences provide a start point for what to do in everyday situations. Beliefs are convictions people hold as true. Values are deep-seated personal beliefs that shape a person's behavior. Values and beliefs are central to character.

4-58. Army leaders should recognize the role beliefs play in preparing Soldiers for battle. Soldiers often fight and win against tremendous odds when they are convinced of the beliefs for which they are fighting. Commitment to such beliefs as justice, liberty, and freedom can be essential ingredients in creating and sustaining the will to fight and prevail. Warrior Ethos is another special case of beliefs.

4-59. Beliefs derive from upbringing, culture, religious backgrounds, and traditions. As a result, different moral beliefs have, and will, continue to be shaped by diverse religious and philosophical traditions. Army leaders serve a Nation that protects the fundamental principle that people are free to choose their own beliefs. America's strength derives and benefits from that diversity. Effective leaders are careful not to require their people to violate their beliefs by ordering or encouraging illegal or unethical actions.

4-60. America's Constitution reflects fundamental national principles. One of these principles is the guarantee of freedom of religion. The Army places a high value on the rights of its Soldiers to observe tenets of their respective religious faiths while respecting individual differences in moral background and personal conviction. While religious beliefs and practices remain a decision of individual conscience, Army leaders are responsible for ensuring their Soldiers and civilians have the opportunity to practice their religion. Commanders, in accordance with regulatory guidance, normally approve requests for accommodation of religious practices unless they will have an adverse impact on unit readiness, individual readiness, unit cohesion, morale, discipline, safety, and/or health. At the same time, no leader may apply undue influence, coerce, or harass subordinates with reference to matters of religion. Chaplains are staff officers with specialized training and specific responsibilities for ensuring the free exercise of religion and are available to advise and assist Army leaders at every level.

4-61. A common theme expressed by American prisoners of war during the Korean and Vietnam wars was the importance of beliefs instilled by a common American culture. Those beliefs helped them to withstand torture and the hardships of captivity.

He Never Gave In

In a park in Alexandria, Virginia is the life size statue of an American Soldier with two small Vietnamese children. Near them is a wall with the names of 65 other Alexandrians who died during the Vietnam conflict.

This memorial came almost forty years after CPT Humbert "Rocky" Versace, a prisoner of war, was executed by his captors in North Vietnam. It honors a man who never gave up his beliefs during extreme hardships and never gave in to the enemy, even in the face of death.

CPT Versace was a West Point graduate assigned to the military assistance advisory group as an intelligence advisor during October 1963.

While accompanying a Civilian Irregular Defense Group engaged in combat operations in the An Xuyen Province, Versace and two fellow Special Forces Soldiers, LT Nick Rowe and SFC Dan Pitzer, were attacked by a Viet Cong main force battalion. Versace, shot in the leg and back, was taken prisoner along with the others.

They were forced to walk barefoot a long distance, deep into the jungle. Once there, Versace assumed the position of senior prisoner and demanded the captors treat them as prisoners, not war criminals. They locked him in an isolation box, beaten and interrogated. He tried to escape four times, once crawling through the surrounding swamp until he was recaptured. He garnered most of the attention of the Viet Cong so that life was tolerable for his fellow prisoners. He was their role model.

He refused to violate the Code of Conduct, giving the enemy only information required by the Geneva Convention which he would recite repeatedly, chapter and verse.

When other Soldiers would operate in those remote areas, they heard stories of Versace's ordeal from local rice farmers. Versace spoke fluent Vietnamese and French and would resist his captors loudly enough that local villagers could hear him. They reported seeing him led through the area bare footed, with a rope around his neck, hands tied, and head swollen and yellow from jaundice. His hair had turned white from the physical stress. The rice farmers spoke of his strength and character and his commitment to his God and his country.

On 26 September 26 1965, after two years in captivity, he was executed in retaliation for three Viet Cong killed in Da Nang. For his bravery, Versace was awarded the Medal of Honor and inducted into the Ranger Hall of Fame at Fort Benning.

Versace's remains were never found, but a tombstone bearing his name stands above an empty grave in Arlington cemetery. The statue across town is a tribute to who Captain Versace was. Ironically, he was just weeks from leaving the Army and studying to become a missionary before being captured. He wanted to return to Vietnam and help the orphaned children. Most of all, he will be remembered as someone with strong character and beliefs who never gave in.

CHARACTER AND ETHICS

4-62. Adhering to the principles that the Army Values embody is essential to upholding high ethical standards of behavior. Unethical behavior quickly destroys organizational morale and cohesion—it undermines the trust and confidence essential to teamwork and mission accomplishment. Consistently doing the right thing forges strong character in individuals and expands to create a culture of trust throughout the organization.

4-63. Ethics are concerned with how a person should behave. Values represent the beliefs that a person has. The seven Army Values represent a set of common beliefs that leaders are expected to uphold and reinforce by their actions. The translation from desirable ethics to internal values to actual behavior involves choices.

4-64. Ethical conduct must reflect genuine values and beliefs. Soldiers and Army civilians adhere to the Army Values because they want to live ethically and profess the values because they know what is right. Adopting good values and making ethical choices are essential to produce leaders of character.

4-65. In combat, ethical choices are not always easy. The right thing may not only be unpopular, but dangerous as well. Complex and dangerous situations often reveal who is a leader of character and who is not. Consider the actions of Warrant Officer Thompson at My Lai, Vietnam.

Warrant Officer Thompson at My Lai, Vietnam

On 16 March 1968, WO1 Hugh C. Thompson, Jr. and his two-man helicopter crew were on a reconnaissance mission over the village of My Lai, Republic of Vietnam. WO1 Thompson watched in horror as he saw an American Soldier shoot an injured Vietnamese child. Minutes later, he observed more Soldiers advancing on a number of civilians in a ditch. Suspecting possible reprisal shootings, WO1 Thompson landed his helicopter and questioned a young officer about what was happening. Told that the ground combat action was none of his business, WO1 Thompson took off and continued to circle the embattled area.

When it became apparent to Thompson that the American troops had now begun firing on more unarmed civilians, he landed his helicopter between the Soldiers and a group of ten villagers headed towards a homemade bomb shelter. Thompson ordered his gunner to train his weapon on the approaching Soldiers and to fire if necessary. Then he personally coaxed the civilians out of the shelter and airlifted them to safety.

WO1 Thompson's immediate radio reports about what was happening triggered a cease-fire order that ultimately saved the lives of many more villagers. Thompson's willingness to place himself in physical danger to do the ethically and morally right thing was a sterling example of personal and moral courage.

4-66. WO1 Thompson's choices prevented further atrocities on the ground and demonstrated that duty-conscious Americans ultimately enforce moral standards of decency. Soldiers must have the personal and moral courage to block criminal behavior and to protect noncombatants.

4-67. Army leaders must consistently focus on shaping ethics-based organizational climates in which subordinates and organizations can achieve their full potential. To reach their goal, leaders can use tools such as the Ethical Climate Assessment Survey (GTA 22-6-1) to assess ethical aspects of their own character and actions, the workplace, and the external environment. Once they have done a climate assessment, leaders prepare and follow a plan of action. The plan of action focuses on solving ethical problems within the leader's span of influence, while the higher headquarters is informed of ethical problems that cannot be changed at the subordinate unit's level.

ETHICAL REASONING

4-68. To be an ethical leader requires more than knowing Army's values. Leaders must be able to apply them to find moral solutions to diverse problems. Ethical reasoning occurs both as an informal process natural to thinking and as an integral part of the formal Army problem solving model (described in Chapter 2, FM 5-0). Ethical considerations occur naturally during all steps of the formal process from identifying the problem through making and implementing the decision. The model specifically states that ethics are explicit considerations when selecting screening criteria, when conducting analysis, and during the comparison of possible solutions.

4-69. Ethical choices may be between right and wrong, shades of gray, or two rights. Some problems center on an ethical dilemma requiring special consideration of what is most ethical. Leaders use multiple perspectives to think about an ethical problem, applying all three perspectives to determine the most ethical choice. One perspective comes from the view that desirable virtues such as courage, justice, and benevolence define ethical outcomes. A second perspective comes from the set of agreed-upon values or rules, such as the Army Values or rights established by the Constitution. A third perspective bases the consequences of the decision on whatever produces the greatest good for the greatest number is most favorable.

4-70. True to the oath they take, Army leaders are expected do the right things for the right reasons all the time. That is why followers count on their leaders to be more than just technically and tactically proficient. They rely on them to make good decisions that are also ethical. Determining what is right and ethical can be a difficult task.

4-71. Ethical dilemmas are nothing new for military leaders. Although it often seems critical to gain timely and valuable intelligence from insurgent detainees or enemy prisoners, what measures are appropriate to obtain vital information from the enemy that could save lives? Vaguely understood instructions from higher headquarters could present one reason why subordinates sometimes push the limits past the framework of what is legal, believing they are doing their duty. Nothing could be more dangerous from an ethical perspective, and nothing could do more harm to the reputation of the Army and its mission. If legal limits are clearly in question, the Army Values bind everyone involved, regardless of rank, to do something about it. Army leaders have a responsibility and the duty to research relevant orders, rules, and regulations and to demand clarification of orders that could lead to criminal misinterpretation or abuse. Ultimately, Army leaders must accept the consequences of their actions.

4-72. Keep in mind that ethical reasoning is very complex in practice. The process to resolve ethical dilemmas involves critical thinking based on the Army Values. No formula will work every time. By embracing the Army Values to govern personal actions, understanding regulations and orders, learning from experiences, and applying multiple perspectives of ethics, leaders will be prepared to face tough calls in life.

ETHICAL ORDERS

4-73. Making the right choice and acting on it when faced with an ethical question can be difficult. Sometimes it means standing firm and disagreeing with the boss on ethical grounds. These occasions test character. Situations in which a leader thinks an illegal order is issued can be the most difficult.

4-74. Under normal circumstances, a leader executes a superior leader's decision with energy and enthusiasm. The only exception would be illegal orders, which a leader has a duty to disobey. If a Soldier perceives that an order is illegal, that Soldier should be sure the details of the order and its original intent are fully understood. The Soldier should seek immediate clarification from the person who gave it before proceeding.

4-75. If the question is more complex, seek legal counsel. If it requires an immediate decision, as may happen in the heat of combat, make the best judgment possible based on the Army Values, personal experience, critical thinking, and previous study and reflection. There is a risk when a leader disobeys what may be an illegal order, and it may be the most difficult decision that Soldier ever makes. Nonetheless, that is what competent, confident, and ethical leaders should do.

4-76. While a leader may not be completely prepared for the complex situations, spending time to reflect on the Army Values, studying, and honing personal leadership competencies will help. Talk to superiors, particularly those who have done the same.

4-77. Living the Army Values and acting ethically is not just for generals and colonels. There are ethical decisions made every day in military units and in offices on Army installations across the world. They include decisions that can directly affect the lives of Soldiers in the field, innocent noncombatants, Army civilians, as well as American taxpayers. It is up to all Army leaders to make value-based, ethical choices for the good of the Army and the Nation. Army leaders should have the strength of character to make the right choices.

Chapter 5
Leader Presence

...[L]eadership is not a natural trait, something inherited like the color of eyes or hair. Actually, leadership is a skill that can be studied, learned, and perfected by practice.

The Noncom's Guide (1962)

5-1. The impression that a leader makes on others contributes to the success in leading them. How others perceive a leader depends on the leader's outward appearance, demeanor, actions, and words.

5-2. Followers need a way to size up their leaders, dependent on leaders being where Soldiers and civilians are. Organizational and strategic level leaders who are willing to go everywhere, including where the conditions are the most severe, illustrate through their presence that they care. There is no greater inspiration than leaders who routinely share in team hardships and dangers. Moving to where duties are performed allows the leader to have firsthand knowledge of the real conditions Soldiers and civilians face. Soldiers and civilians who see or hear from the boss appreciate knowing that their unit has an important part to play.

5-3. Presence is not just a matter of the leader showing up; it involves the image that the leader projects. Presence is conveyed through actions, words, and the manner in which leaders carry themselves. A reputation is conveyed by the respect that others show, how they refer to the leader, and respond to the leader's guidance. Presence is a critical attribute that leaders need to understand. A leader's effectiveness is dramatically enhanced by understanding and developing the following areas:

- **Military bearing**: projecting a commanding presence, a professional image of authority.
- **Physical fitness:** having sound health, strength, and endurance, which sustain emotional health and conceptual abilities under prolonged stress.
- **Confidence**: projecting self-confidence and certainty in the unit's ability to succeed in whatever it does; able to demonstrate composure and outward calm through steady control over emotion.
- **Resilience**: showing a tendency to recover quickly from setbacks, shock, injuries, adversity, and stress while maintaining a mission and organizational focus.

5-4. Physical characteristics—military and professional bearing, health and physical fitness—can and must be continuously developed in order to establish presence. Army leaders represent the institution and government and should always maintain an appropriate level of physical fitness and professional bearing.

MILITARY AND PROFESSIONAL BEARING

Our quality soldiers should look as good as they are.

Julius W. Gates
Sergeant Major of the Army (1987-1991)

5-5. Pride in self starts with pride in appearance. Army leaders are expected to look and act like professionals. They must know how to wear the appropriate uniform or civilian attire and do so with pride. Soldiers seen in public with their jackets unbuttoned and ties undone do not send a message of pride and professionalism. Instead, they let down their unit and fellow Soldiers in the eyes of the American people. Meeting prescribed height and weight standards is another integral part of the professional role. How leaders carry themselves when displaying military courtesy and appearance sends a clear signal: I am proud of my uniform, my unit, and my country.

5-6. Skillful use of professional bearing—fitness, courtesy, and proper military appearance—can also aid in overcoming difficult situations. A professional presents a decent appearance because it commands respect. Professionals must be competent as well. They look good because they are good.

HEALTH FITNESS

5-7. Disease remains a potent enemy on modern battlefields. Staying healthy and physically fit is important to protect Soldiers from disease and strengthen them to deal with the psychological impact of combat. A Soldier is similar to a complex combat system. Just as a tank requires good maintenance and fuel at regular intervals, a Soldier needs exercise, sufficient sleep, and adequate food and water for peak performance.

5-8. Health fitness is everything done to maintain good health. It includes undergoing routine physical exams; practicing good dental hygiene, personal grooming, and cleanliness; keeping immunizations current; as well as considering mental stresses. Healthy and hygiene-conscious Soldiers perform better in extreme operational environments. One sick crewmember on a well-trained flight crew represents a weak link in the chain and makes the entire aircraft more vulnerable and less lethal. Health fitness also includes avoiding things that can degrade personal health, such as substance abuse, obesity, and smoking.

PHYSICAL FITNESS

...I am obliged to sweat them tonight, sir, so that I can save their blood tomorrow.

Thomas J. "Stonewall" Jackson
Confederate Civil War General (1861-1863)

5-9. Unit readiness begins with physically fit Soldiers and leaders, for combat drains physically, mentally, and emotionally. Physical fitness, while crucial for success in battle, is important for all members of the Army team, not just Soldiers. Physically fit people feel more competent and confident, handle stress better, work longer and harder, and recover faster. These attributes provide valuable payoffs in any environment.

5-10. The physical demands of leadership, prolonged deployments, and continuous operations can erode more than physical attributes. Physical fitness and adequate rest support cognitive functioning and emotional stability, both essential for sound leadership. Soldiers must be prepared for deprivation; it is difficult to maintain high levels of fitness during fast-paced, demanding operations. If not physically fit before deployment, the effects of additional stress compromise mental and emotional fitness as well. Combat operations in difficult terrain, extreme climates, and high altitude require extensive physical pre-conditioning; once in the area of operations there must be continued efforts to sustain physical readiness.

5-11. Preparedness for operational missions must be a primary focus of the unit's physical fitness program. Fitness programs that merely emphasize top scores on the Army physical fitness test do not prepare Soldiers for the strenuous demands of actual combat. The forward-looking leader develops a balanced physical fitness program that enables Soldiers to execute the unit's mission-essential task list. (FM 7-0 discusses the integration of Soldier, leader, and collective training based on the mission-essential task list).

5-12. Ultimately, the physical fitness requirements for Army leaders have significant impact on their personal performance and health. Since leaders' decisions affect their organizations' combat effectiveness, health, and safety, it is an ethical as well as a practical imperative for leaders to remain healthy and fit.

CONFIDENCE

5-13. Confidence is the faith that leaders place in their abilities to act properly in any situation, even under stress and with little information. Leaders who know their own capabilities and believe in themselves are confident. Self-confidence grows from professional competence. Too much confidence can be as detrimental as too little confidence. Both extremes impede learning and adaptability. Bluster—loudmouthed bragging or self-promotion—is not confidence. Truly confident leaders do not need to advertise their gift because their actions prove their abilities.

5-14. Confidence is important for leaders and teams. The confidence of a good leader is contagious and quickly permeates the entire organization, especially in dire situations. In combat, confident leaders help Soldiers control doubt while reducing team anxiety. Combined with strong will and self-discipline, confidence spurs leaders to do what must be done in circumstances where it would be easier to do nothing.

RESILIENCE

5-15. Resilient leaders can recover quickly from setbacks, shock, injuries, adversity, and stress while maintaining their mission and organizational focus. Their resilience rests on will, the inner drive that compels them to keep going, even when exhausted, hungry, afraid, cold, and wet. Resilience helps leaders and their organizations to carry difficult missions to their conclusion.

5-16. Resilience and the will to succeed are not sufficient to carry the day during adversity. Competence and knowledge guide the energies of a strong will to pursue courses of action that lead to success and victory in battle. The leader's premier task is to instill resilience and a winning spirit in subordinates. That begins with tough and realistic training.

5-17. Resilience is essential when pursuing mission accomplishment. No matter what the working conditions are, a strong personal attitude helps prevail over any adverse external conditions. All members of the Army—active, reserve, or civilian—will experience situations when it would seem easier to quit rather than finish the task. During those times, everyone needs an inner source of energy to press on to mission completion. When things go badly, a leader must draw on inner reserves to persevere.

5-18. The following story of a U.S. military police company in action illustrates how individuals and leaders showed resilience and discipline when faced with the shock of an ambush by a superior number of insurgents during routine convoy operations.

Mission First—Never Quit!

When SGT Leigh Ann Hester and members of her Kentucky National Guard military police company set out for a routine convoy escort mission in March 2005, she did not know what challenges awaited her and her team.

SGT Hester was the vehicle commander riding in the second HMMWV behind a convoy of 26 supply vehicles when her squad leader, SSG Timothy Nein, observed the convoy under attack and moved to contact.

When she arrived at the ambush location, she saw the lead vehicle had been hit with a rocket-propelled grenade. A group of about 50 insurgents seemed determined to inflict devastating damage on the now stopped convoy. She immediately joined the fight and engaged the enemy with well-aimed fires from her rifle and grenade launcher. The intense engagement lasted over 45 minutes. When the firing finally subsided, 27 insurgents lay dead, six were wounded, and one was captured.

Despite the initially overwhelming odds and battlefield clutter, SGT Hester and her Soldiers persevered. They effectively quelled the attack, allowing the supply convoy to continue safely to their destination. Throughout the situational chaos, SGT Hester and her comrades had remained resilient, focused, and professional. The fearless response by Hester and SSG Nein had helped the Soldiers overcome the initial shock of the ambush and instilled the necessary confidence and courage to complete the mission successfully.

For Hester's military police company, the countless hours on small arms ranges and practicing urban warfare and convoy operations had paid off.

Well-rehearsed battle drills became second nature. She and her fellow Soldiers were able to live the words, "Mission first—never quit."

For her actions, SGT Hester earned the Silver Star. She is the first female Soldier since World War II to receive this award. SSG Nein and SPC Jason Mike also won the Silver Star; several other unit members were awarded Bronze Stars for valor.

Chapter 6
Leader Intelligence

6-1. An Army leader's intelligence draws on the mental tendencies and resources that shape conceptual abilities, which are applied to one's duties and responsibilities. Conceptual abilities enable sound judgment before implementing concepts and plans. They help one think creatively and reason analytically, critically, ethically, and with cultural sensitivity to consider unintended as well as intended consequences. Like a chess player trying to anticipate an opponent's moves three or four turns in advance (action-reaction-counteraction), leaders must think through what they expect to occur because of a decision. Some decisions may set off a chain of events. Therefore, leaders must attempt to anticipate the second- and third-order effects of their actions. Even lower-level leaders' actions may have effects well beyond what they expect.

6-2. The conceptual components affecting the Army leader's intelligence include—

- Agility.
- Judgment.
- Innovation.
- Interpersonal tact.
- Domain knowledge.

MENTAL AGILITY

It is not genius which reveals to me suddenly and secretly what I should do in circumstances unexpected by others; it is thought and meditation.

Napoleon Bonaparte
French general (1789-1804) and Emperor of France (1804-1814)

6-3. Mental agility is a flexibility of mind, a tendency to anticipate or adapt to uncertain or changing situations. Agility assists thinking through second- and third-order effects when current decisions or actions are not producing the desired effects. It helps break from habitual thought patterns, to improvise when faced with conceptual impasses, and quickly apply multiple perspectives to consider new approaches or solutions.

6-4. Mental agility is important in military leadership because great militaries adapt to fight the enemy, not the plan. Agile leaders stay ahead of changing environments and incomplete planning to preempt problems. In the operational sense, agility also shows in the ability to create ad hoc and tactically creative units that adapt to changing situations. They can alter their behavior to ease transitioning from full-scale maneuver war to stability operations in urban areas.

6-5. The basis for mental agility is the ability to reason critically while keeping an open mind to multiple possibilities until reaching the most sensible solution. Critical thinking is a thought process that aims to find truth in situations where direct observation is insufficient, impossible, or impractical. It allows thinking through and solving problems and is central to decision making. Critical thinking is the key to understanding changing situations, finding causes, arriving at justifiable conclusions, making good judgments, and learning from experience.

6-6. Critical thinking implies examining a problem in depth, from multiple points of view, and not settling for the first answer that comes to mind. Army leaders need this ability because many of the choices they face require more than one solution. The first and most important step in finding an appropriate solution is to isolate the main problem. Sometimes determining the real problem presents a huge hurdle; at other times, one has to sort through distracting multiple problems to get to the real issue.

6-7. A leader's mental agility in quickly isolating a problem and identifying solutions allows the use of initiative to adjust to change during operations. Agility and initiative do not appear magically. The leader must instill them within all subordinates by creating a climate that encourages team participation. Identifying honest mistakes in training makes subordinates more likely to develop their own initiative.

6-8. Modern Army training and education focuses on improving leader agility and small unit initiative. Combat deployments in Grenada, Panama, Kosovo, Somalia, Afghanistan, and Iraq have emphasized the demands on mental agility and tactical initiative down to the level of the individual Soldier. Contemporary operational environments call for more agile junior officers and noncommissioned officers, able to lead effectively small and versatile units across the spectrum of conflicts.

SOUND JUDGMENT

Judgment comes from experience and experience comes from bad judgments.

General of the Army Omar N. Bradley
Address at the U.S. Army War College (1971)

6-9. Judgment goes hand in hand with agility. Judgment requires having a capacity to assess situations or circumstances shrewdly and to draw feasible conclusions. Good judgment enables the leader to form sound opinions and to make sensible decisions and reliable guesses. Good judgment on a consistent basis is important for successful Army leaders and much of it comes from experience. Leaders acquire experience through trial and error and by watching the experiences of others. Learning from others can occur through mentoring and coaching by superiors, peers, and even some subordinates (see Part Three for more information). Another method of expanding experience is self-development by reading biographies and autobiographies of notable men and women to learn from their successes and failures. The histories of successful people offer ageless insights, wisdom, and methods that might be adaptable to the current environment or situation.

6-10. Often, leaders must juggle facts, questionable data, and gut-level feelings to arrive at a quality decision. Good judgment helps to make the best decision for the situation at hand. It is a key attribute of the art of command and the transformation of knowledge into understanding and quality execution. FM 6-0 discusses how leaders convert data and information into knowledge and understanding.

6-11. Good judgment contributes to an ability to determine possible courses of action and decide what action to take. Before choosing the course of action, consider the consequences and think methodically. Some sources that aid judgment are senior leaders' intents, the desired outcome, rules, laws, regulations, experience, and values. Good judgment includes the ability to size up subordinates, peers, and the enemy for strengths, weaknesses, and to create appropriate solutions and action. Like agility, it is a critical part of problem solving and decision making.

INNOVATION

6-12. Innovation describes the Army leader's ability to introduce something new for the first time when needed or an opportunity exists. Being innovative includes creativity in the production of ideas that are original and worthwhile.

6-13. Sometimes a new problem presents itself or an old problem requires a new solution. Army leaders should seize such opportunities to think creatively and to innovate. The key concept for creative thinking is developing new ideas and ways to challenge subordinates with new approaches and ideas. It also involves devising new ways for their Soldiers and civilians to accomplish tasks and missions. Creative thinking includes using adaptive approaches (drawing from previous similar circumstances) or innovative approaches (coming up with a completely new idea).

6-14. All leaders can and must think creatively to adapt to new environments. A unit deployed for stability operations may find itself isolated on a small secure compound with limited athletic facilities and without much room to run. This situation would require its leaders to devise reliable ways to maintain their Soldiers' physical fitness. Innovative solutions might include weight training, games, stationary runs, aerobics, treadmills, and other fitness drills.

6-15. Innovative leaders prevent complacency by finding new ways to challenge subordinates with forward-looking approaches and ideas. To be innovators, leaders learn to rely on intuition, experience, knowledge, and input from subordinates. Innovative leaders reinforce team building by making everybody responsible for, and stakeholders in, the innovation process.

INTERPERSONAL TACT

6-16. Effectively interacting with others depends on knowing what others perceive. It also relies on accepting the character, reactions, and motives of oneself and others. Interpersonal tact combines these skills, along with recognizing diversity and displaying self-control, balance, and stability in all situations.

RECOGNIZING DIVERSITY

6-17. Soldiers, civilians, and contractors originate from vastly different backgrounds and are shaped by schooling, race, gender, religion, as well as a host of other influences. Personal perspectives can even vary within societal groups. People should avoid snap conclusions based on stereotypes. It is better to understand individuals by acknowledging their differences, qualifications, contributions, and potential.

6-18. Joining the Army as Soldiers and civilians, subordinates agreed to accept the Army's culture. This initial bond holds them together. Army leaders further strengthen the team effort by creating an environment where subordinates know they are valued for their talents, contributions, and differences. A leader's job is not to make everyone the same; it is to take advantage of the different capabilities and talents brought to the team. The biggest challenge is to put each member in the right place to build the best possible team.

6-19. Army leaders should keep an open mind about cultural diversity. It is important, because it is unknown how the talents of certain individuals or groups will contribute to mission accomplishment. During World War II, U.S. Marines from the Navajo nation formed a group of radio communications specialists called the Navajo Code Talkers. The code talkers used their native language—a unique talent—to handle command radio traffic. Using the Navajo code significantly contributed to successful ground operations because the best Japanese code breakers could not decipher their messages.

SELF-CONTROL

...[A]n officer or noncommissioned officer who loses his temper and flies into a tantrum has failed to obtain his first triumph in discipline.

Noncommissioned Officer's Manual (1917)

6-20. Good leaders control their emotions. Instead of hysterics or showing no emotion at all, leaders should display the right amount of sensitivity and passion to tap into subordinates' emotions. Maintaining self-control inspires calm confidence in the team. Self-control encourages feedback from subordinates that can expand understanding of what is really happening. Self-control in combat is especially important for Army leaders. Leaders who lose their self-control cannot expect those who follow them to maintain theirs.

Self-Control

A leader's emotional state is often transferred to subordinates. A battalion staff team at the National Training Center demonstrates how short tempers, fatigue, and stress can have a devastating effect.

During the first week of force-on-force operations MAJ Jones* had been under a lot of stress and had gotten little sleep. MAJ Jones had earned a reputation for a short temper, but nothing prepared the staff for what happened next.

MAJ Jones had snapped, causing a commotion outside: "You need to get your lieutenants under control; I let that idiot Smith use my HMMWV but told him to be back here by 1400. Now I'll miss the brigade rehearsal. Who does he think he is?"

A fellow major tried to calm the situation down by offering to give MAJ Jones a ride to the rehearsal. "No! I want MY HMMWV! When that idiot gets back I want him standing right here," kicking his heel into the desert sand. "No food, no water, he better be waiting for me when I get back."

The "idiot" had been at the brigade headquarters picking up the next operation order so the staff could start mission planning. He had also stopped to re-fuel MAJ Jones's HMMWV so he would not run out of fuel on the way to the rehearsal. He was doing his job.

MAJ Jones was obviously overstressed by the situation, his job, and the demanding pace of the operations, just like everyone else on the battalion staff. His failure to control his anger and maintain his professional bearing cost him the respect and loyalty of many of his fellow officers and enlisted Soldiers. It also planted a seed of doubt about how he would perform under real combat conditions.

Leaders do not have the luxury of being able to lose their temper, be unprofessional, or berate subordinate leaders. Every action is noticed and although some Soldiers dismissed the incident, some carried the memory throughout their career. MAJ Jones's blunder served Soldiers differently that day. Some saw the effects of stress and its impacts and some saw something they never wanted to become.

*names have been changed

EMOTIONAL FACTORS

...[A]nyone can get angry—that is easy... but to [get angry with] the right person, to the right extent, at the right time, for the right reason, and in the right way is no longer something easy that anyone can do.

Aristotle
Greek philosopher and tutor to Alexander the Great

6-21. An Army leader's self-control, balance, and stability greatly influence his ability to interact with others. People are human beings with hopes, fears, concerns, and dreams. Understanding that motivation and endurance are sparked by emotional energy is a powerful leadership tool. Giving constructive feedback will help mobilize the team's emotional energies to accomplish difficult missions during tough times.

6-22. Self-control, balance, and stability also assist making the right ethical choices. (Chapter 4 covers ethical reasoning.) An ethical leader successfully applies ethical principles to decision making and retains self-control. Leaders cannot be at the mercy of emotion. It is critical for leaders to remain calm under pressure and expend energy on things they can positively influence and not worry about things they cannot affect.

6-23. Emotionally mature and competent leaders are also aware of their own strengths and weaknesses. They spend their energy on self-improvement, while immature leaders usually waste their energy denying that there is anything wrong or analyzing the shortcomings of others. Mature, less defensive leaders benefit from feedback in ways that immature people cannot.

BALANCE

6-24. Emotionally balanced leaders are able to display the right emotion for a given situation and can read others' emotional state. They draw on their experience and provide their subordinates the proper perspective on unfolding events. They have a range of attitudes, from relaxed to intense, with which to approach diverse situations. They know how to choose the one appropriate for the circumstances. Balanced leaders know how to convey that things are urgent without throwing the entire organization into chaos. They are able to encourage their people to continue the mission, even in the toughest of moments.

STABILITY

6-25. Effective leaders are steady, levelheaded when under pressure and fatigued, and calm in the face of danger. These characteristics stabilize their subordinates who are always looking to their leader's example:

- Model the emotions for subordinates to display.
- Do not give in to the temptation to do what personally feels good.
- If under great stress, it might feel better to vent—but will that help the organization?
- If subordinates are to be calm and rational under pressure, leaders must display the same stability.

6-26. Brigadier General Thomas J. Jackson's actions during the Civil War's First Battle of Bull Run serve as a vivid example of how one leader's self-control under fire can stabilize an uncertain situation and ultimately turn the tide in battle.

He Stood Like a Stone Wall

At a turning point during the famous First Battle of Bull Run, the Confederates were down for the count: 20,000 Union soldiers were encroaching upon the position of the mere 1,000 Confederate men led by Colonel Nathan Evans. Despite the arrival of reinforcements led by Brigadier General Barnard E. Bee and Colonel Francis Bartow, much of the Confederate line received a devastating surprise attack at their Matthews Hill positions, and had to retreat.

Only then, at this seemingly hopeless moment of battle, did a certain Brigadier General Thomas J. Jackson and his 2,000-man brigade ride in to protect the remaining Confederate troops. BG Jackson set up his defensive line on Henry Hill, leading to a ruthless artillery duel between BG Jackson's thirteen 6-pounder guns and the Union's eleven, just 300 yards away.

Despite the fact that all the Confederate troops were surrendering and fleeing past him, BG Jackson stood still. BG Bee said in a panic, "The enemy are driving us!" Jackson, calm, composed, and determined, replied with the statement, "Then, Sir, we will give them the bayonet."

As the story goes, BG Bee took BG Jackson's words as inspiration. He rode back to his own troops, ordering them to reform and fight on. In the midst of this gory battle, he exclaimed to his troops, "There is Jackson standing like a stone wall. Let us determine to die here, and we will conquer. Rally behind the Virginians."

Despite the fact that BG Bee was fatally wounded and died the next day, the Confederate line held. BG "Stonewall" Jackson kept the nickname that BG Bee had given him, and is now recognized as a staple of American heroism and calm bravery in the face of seemingly certain destruction.

DOMAIN KNOWLEDGE

6-27. Domain knowledge requires possessing facts, beliefs, and logical assumptions in many areas. Tactical knowledge is an understanding of military tactics related to securing a designated objective through military means. Technical knowledge consists of the specialized information associated with a particular function or system. Joint knowledge is an understanding of joint organizations, their procedures, and their roles in national defense. Cultural and geopolitical knowledge is awareness of cultural, geographic, and political differences and sensitivities.

TACTICAL KNOWLEDGE

> *The commander must decide how he will fight the battle* before it begins. *He must then decide how he will use the military effort at his disposal to force the battle to swing the way he wishes it to go; he must make the enemy dance to his tune from the beginning, and never vice versa.*
>
> Field Marshal Viscount Montgomery
> *Memoirs* (1958)

Doctrine

6-28. Army leaders know doctrine, tactics, techniques, and procedures. Their tactical knowledge allows them to effectively employ individuals, teams, and larger organizations together with the activities of systems (combat multipliers) to fight and win engagements and battles or to achieve other objectives. While direct leaders usually fight current battles, organizational leaders focus deeper in time, space, and events. This includes a geopolitical dimension.

6-29. Tactics is the art and science of employing available means to win battles and engagements. The science of tactics encompasses capabilities, techniques, and procedures that can be codified. The art includes the creative and flexible array of means to accomplish assigned missions, decision making when facing an intelligent enemy, and the effects of combat on Soldiers. FM 3-90 addresses tactics. FM 71-100 addresses divisional organizations, tactics, and techniques. FM 100-15 contains the same information for corps. FM 100-7 discusses the Army in theater operations.

Fieldcraft

6-30. Fieldcraft describes the skills Soldiers require to sustain themselves in the field. Proficiency in fieldcraft reduces the likelihood of casualties. Understanding and excelling at fieldcraft sets conditions for mission success. Likewise, the requirement that Army leaders make sure their Soldiers take care of themselves and provide them with the means to do so also sets conditions for success.

6-31. STP 21-1-SMCT, *Soldier's Manual of Common Tasks*, lists the individual skills all Soldiers must know to operate effectively in the field. Those skills include everything from staying healthy to digging fighting positions. Some military occupational specialties require proficiency in additional fieldcraft skills. They are listed in Soldiers' manuals for these specialties.

6-32. Army leaders gain proficiency in fieldcraft through formal training, study, and practice. Although easily learned, fieldcraft skills are often neglected during training exercises. That is why during peacetime exercises, leaders must strictly enforce tactical discipline and make sure their Soldiers practice fieldcraft to keep them from becoming casualties in wartime. The Army's Combat Training Centers set the right example on how to conduct realistic training in an environment that enforces tactical and fieldcraft discipline. During Combat Training Center rotations, skilled observers and controllers assess appropriate training casualties and make recommendations to reinforce the appropriate fieldcraft standards.

Tactical Proficiency

6-33. While practicing tactical abilities is generally challenging, competent leaders try to replicate actual operational conditions during battle-focused training (see FM 7-0). Unfortunately, Army leaders cannot always take their entire unit to the field for full-scale maneuvers. They must therefore learn to achieve maximum readiness by training parts of a scenario or a unit on the ground, while exercising larger echelons with simulations. Despite distracters and limitations, readiness-focused leaders train for war as realistically as possible. FM 7-0 and FM 7-1 discuss training principles and techniques.

Technical Knowledge

Knowing Equipment

6-34. Technical knowledge relates to equipment, weapons, and systems—everything from a gun sight to the computer that tracks personnel actions. Since direct leaders are closer to their equipment than organizational and strategic leaders, they have a greater need to know how it works and how to use it. Direct leaders are usually the experts called upon to solve problems with equipment. They figure out how to make it work better, how to apply it, how to fix it, and even how to modify it. If they do not know the specifics, they will know who knows how to solve issues with it. Subordinates expect their first-line leaders to know the equipment and be experts in all the applicable technical skills. That is why sergeants, junior officers, warrant officers, wage grade employees, and journeymen are the Army's technical experts and teachers.

Operating Equipment

6-35. Military and civilian leaders know how to operate their organizations' equipment and ensure their people do as well. They often set an example with a hands-on approach. When new equipment arrives, direct leaders learn how to use it and train their subordinates to do the same. Once individuals are trained, teams, and in turn, whole units train together. Army leaders know understanding equipment strengths and weaknesses are critical. Adapting to these factors is necessary to achieve success in combat.

Employing Equipment

6-36. Direct, organizational, and strategic level leaders need to know what functional value the equipment has for their operations and how to employ the equipment in their units and organizations. At higher levels, the requirement for technical knowledge shifts from understanding how to operate single items of equipment to how to employ entire systems. Higher-level leaders have a responsibility to keep alert to future capabilities and the impact that fielding will have on their organizations. Some organizational and strategic level leaders have general oversight responsibility for the development of new systems; they should have knowledge of the major features and required capabilities. Their interests are in knowing the technical aspects of how systems affect doctrine, organizational design, training, related materiel, personnel, and facilities. They must ensure that organizations are provided with all necessary resources to properly field, train, maintain, operate, inventory, and turn-in equipment.

Joint Knowledge

6-37. Joint warfare is team warfare. The 1986 Goldwater-Nichols legislation mandated a higher level of cooperation among America's military Services, based on experiences drawn from previous deployments. Since then, Army leaders from the most junior field leader to the generals serving at the strategic level have embraced the importance of joint warfare. Leaders acquire joint knowledge through formal training in the Joint Professional Military Education program and assignments in joint organizations and staffs. Army leaders acknowledge all Services bring certain strengths and limitations to the battlefield. Only the close cooperation of all Services can assure swift mission accomplishment in the complex operational environments our militaries face.

Cultural and Geopolitical Knowledge

> *If you can wear Arab kit when with the tribes you will acquire their trust and intimacy to a degree impossible in uniform.*
>
> T.E. Lawrence
> *Twenty-Seven-Articles* (1917)

6-38. Culture consists of shared beliefs, values, and assumptions about what is important. Army leaders are mindful of cultural factors in three contexts:

- Sensitive to the different backgrounds of team members to best leverage their talents.
- Aware of the culture of the country in which the organization is operating.
- Consider and evaluate the possible implications of partners' customs, traditions, doctrinal principles, and operational methods when working with forces of another nation.

6-39. Understanding the culture of adversaries and of the country in which the organization is operating is just as important as understanding the culture of a Soldier's own country and organization. Contemporary operational environments, which place smaller units into more culturally complex situations with continuous media coverage, require even greater cultural and geopolitical awareness from every Army leader. Consequently, be aware of current events—particularly those in areas where America has national interests. Before deploying, ensure that Soldiers and the organization are properly prepared to deal with the population of particular areas—either as partners, neutrals, or adversaries. The more that is known about them, including their language, the better off the organization will be.

6-40. Understanding other cultures applies to full spectrum operations, not only stability and reconstruction operations. For example, different tactics may be employed against an adversary who considers surrender a dishonor worse than death, as compared to those for whom surrender remains an honorable option. Likewise, if the organization is operating as part of a multinational team, how well leaders understand partners' capabilities and limitations will affect how well the team accomplishes its mission.

6-41. Cultural understanding is crucial to the success of multinational operations. Army leaders take the time to learn the customs and traditions as well as the operational procedures and doctrine of their partners. To be able to operate successfully in a multinational setting, U.S. leaders must be aware of any differences in doctrinal terminology and the interpretation of orders and instructions. They must learn how and why others think and act as they do. In multinational forces, effective leaders often create a "third culture" by adopting practices from several cultures to create a common operating basis.

6-42. Besides overcoming language barriers, working in a multicultural environment requires leaders to keep plans and orders as simple as possible to prevent misunderstandings and unnecessary losses. Dedicated liaison teams and linguists provide a cultural bridge between partners to mitigate some differences, but they cannot eliminate all of them. FM 3-16 provides information on working in a multinational context.

6-43. Cultural awareness played a major role in the peaceful capture of Najaf during Operation Iraqi Freedom in April 2003.

No Slack Soldiers Take a Knee

The Soldiers of LTC Christopher Hughes' 2nd Battalion, 327th Infantry were tired following several weeks of battling insurgents on their journey to Najaf. It was early April 2003 and elements from the 101st Airborne Division were taking part in a bigger effort to secure the holy city on the road to Baghdad.

The 2-327th had served in Vietnam and one of their finest had been killed just days before rotating to the states. In his honor, and based on his favorite saying "cut the enemy no slack," the battalion now called themselves "No Slack."

Their leader, LTC Hughes, was no stranger to Muslim customs, learning all he could while investigating the bombing of the USS Cole and serving on a joint antiterrorism task force. Still, he took the opportunity to learn more about the Shiite people and the grand Ali Mosque in the city where he and his Soldiers were headed. Earlier that month, on the 54-hour drive out of Kuwait, Hughes had listened while his Iraqi-American translator explained the importance of the Ayatollah Ali Sistani, the years he spent imprisoned under Saddam Hussein, and how Shiites considered the gold-domed Mosque as a most holy site.

When Hughes and his Soldiers approached the mosque to ask Sistani to issue a fatwa (religious decree) allowing the Americans to go on to Baghdad without resistance, they met an angry crowd.

Hundreds of people protected the entrance to the mosque, concerned that the Americans had come to destroy it. They chanted "In city yes—in city OK. Mosque no!" Hughes had to act quickly to dispel their fears. At first, he pointed his weapon to the ground. No one noticed.

Next, he commanded his troops to take a knee. Some gave him a questioning glance, but still obeyed without hesitation. They trusted their leader. Many Iraqis in the crowd joined them; LTC Hughes went a step further. He told his Soldiers to smile. The Iraqis smiled back. The anger in the crowd was defused. A universal language of goodwill spread, and Hughes was able to have his Soldiers get up and walk away.

As he turned to leave, Hughes put his right hand on his chest in a traditional Islamic gesture, "Peace be with you," he said, "Have a nice day." The fatwa was issued, Baghdad was taken, and unnecessary conflict was avoided.

Understanding the mixture of cultures and with an adaptability that makes the American Soldier unique, these combat-hardened warriors allowed diplomacy and respect for others to rule the day.

6-44. Cultural awareness and geopolitical knowledge are important factors when Army leaders are challenged to extend influence beyond their traditional chain of command. There is more about this important topic in Chapter 7.

PART THREE

Competency-Based Leadership for Direct Through Strategic Levels

In short, Army leaders in this century need to be pentathletes, multi-skilled leaders who can thrive in uncertain and complex operating environments... innovative and adaptive leaders who are expert in the art and science of the profession of arms.

The Army needs leaders who are decisive, innovative, adaptive, culturally astute, effective communicators and dedicated to life-long learning.

Dr. Francis J. Harvey
Secretary of the Army
Speech for U.S. Army Command and General Staff College graduation (2005)

Leaders serve to provide purpose, direction and motivation. Army leaders work hard to **lead** people, to **develop** themselves, their subordinates, and organizations, and to **achieve** mission accomplishment across the spectrum of conflicts.

For leadership to be effective in the operational environment, it is important to consider the impact of its dimensions on the members of the organization. Weather and terrain, combined with the day-night cycle, form the basis for all operations. This basic environment is influenced by technology, affecting the application of firepower, maneuver, protection and leadership. A combination of the psychological impact of mortal danger, weapons effects, difficult terrain, and the presence of enemy forces can create chaos and confusion, turning simple tactical and operational plans into the most challenging endeavors.

Continuously building and refining values and attributes, as well as acquiring professional knowledge, is only part of becoming a competent leader. Leadership succeeds when the leader effectively acts and applies the core leader competencies and their subsets. As one moves from direct leadership positions to the organizational and strategic leader levels, those competencies take on different nuances and complexities.

As a direct leader, an example of leading would be providing mission intent. At the organizational level the leader might provide a vision and empower others, while at the strategic level the same leader would lead change and shape an entire insititution for future success. A more thorough discussion of the challenges of leading across different organizational levels and how the core leader competencies are adapted to meet these challenges is found in Chapters 11 and 12.

Chapter 7
Leading

The American soldier...demands professional competence in his leaders. In battle, he wants to know that the job is going to be done right, with no unnecessary casualties. The noncommissioned officer wearing the chevron is supposed to be the best soldier in the platoon and he is supposed to know how to perform all the duties expected of him. The American soldier expects his sergeant to be able to teach him how to do his job. And he expects even more from his officers.

Omar N. Bradley
General of the Army (1950-1953)

7-1. Army leaders apply character, presence, intellect, and abilities to the core leader competencies while guiding others toward a common goal and mission accomplishment. Direct leaders influence others person-to-person, such as a team leader who instructs, recognizes achievement, and encourages hard work. Organizational and strategic leaders influence those in their sphere of influence, including immediate subordinates and staffs, but often guide their organizations using indirect means of influence. At the direct level, a platoon leader knows what a battalion commander wants done, not because the lieutenant was briefed personally, but because the lieutenant understands the commander's intent two levels up. The intent creates a critical link between the organizational and direct leadership levels. At all levels, leaders take advantage of formal and informal processes (see Chapter 3) to extend influence beyond the traditional chain of command.

7-2. The **leading** category of the core leader competencies includes four competencies. (See Appendix A for descriptions and examples of the core leader competencies.) Two competencies focus on who is being led and with what degree of authority and influence: **leads others** and **extends influence beyond the chain of command**. The other leading competencies address two ways by which leaders to convey influence: **leads by example** and **communicates**.

- **Leads others** involves influencing Soldiers or Army civilians in the leader's unit or organization. This competency has a number of components including setting clear direction, enforcing standards, and balancing the care of followers against mission requirements so they are a productive resource. Leading within an established chain of command with rules, procedures, and norms differs from leading outside an established organization or across commands.
- **Extends influence beyond the chain of command** requires the ability to operate in an environment, encompassing higher and lower command structures, and using one's influence outside the traditional chain of command. This includes connecting with joint, allied, and multinational partners, as well as local nationals, and civilian-led governmental or nongovernmental agencies. In this area, leaders often must operate without designated authority or while their authority is not recognized by others.
- **Leads by example** is essential to leading effectively over the course of time. Whether they intend to or not, leaders provide an example that others consider and use in what they do. This competency reminds every leader to serve as a role model. What leaders do should be grounded in the Army Values and imbued with the Warrior Ethos.
- **Communicates** ensures that leaders attain a clear understanding of what needs to be done and why within their organization. This competency deals with maintaining clear focus on the team's efforts to achieve goals and tasks for mission accomplishment. It helps build consensus and is a critical tool for successful operations in diverse multinational settings. Successful leaders refine their communicating abilities by developing advanced oral, written, and listening

skills. Commanders use clear and concise mission orders and other standard forms of communication to convey their decisions to subordinates.

LEADS OTHERS

7-3. Former Army Chief of Staff Creighton W. Abrams once said,

> *The Army is people; its readiness to fight depends upon the readiness of its people, individually and as units. We improve our readiness and foster a ready state of mind by training, motivating and supporting our people, and by giving them a sense of participation in the Army's important endeavors.*

7-4. All of the Army's core leader competencies, especially leading others, involve influence. Army leaders can draw on a variety of techniques to influence others. These range from obtaining compliance to building a commitment to achieve. Compliance is the act of conforming to a specific requirement or demand. Commitment is willing dedication or allegiance to a cause or organization. Resistance is the opposite of compliance and commitment. There are many techniques for influencing others to comply or commit, and leaders can use one or more of them to fit to the specifics of any situation.

COMPLIANCE AND COMMITMENT

7-5. Compliance-focused influence is based primarily on the leader's authority. Giving a direct order to a follower is one approach to obtain compliance during a task. Compliance is appropriate for short-term, immediate requirements and for situations where little risk can be tolerated. Compliance techniques are also appropriate for leaders to use with others who are relatively unfamiliar with their tasks or unwilling or unable to commit fully to the request. If something needs to be done with little time for delay, and there is not a great need for a subordinate to understand why the request is made, then compliance is an acceptable approach. Compliance-focused influence is not particularly effective when a leader's greatest aim is to create initiative and high esteem within the team.

7-6. Commitment-focused influence generally produces longer lasting and broader effects. Whereas compliance only changes a follower's behavior, commitment reaches deeper—changing attitudes and beliefs, as well as behavior. For example, when a leader builds responsibility among followers, they will likely demonstrate more initiative, personal involvement, and creativity. Commitment grows from an individual's desire to gain a sense of control and develop self-worth by contributing to the organization. Depending on the objective of the influence, leaders can strengthen commitment by reinforcing followers' identification with the Nation (loyalty), the Army (professionalism), the unit or organization (selfless service), the leadership in a unit (respect), and to the job (duty).

Influence Techniques

7-7. Leaders use several specific techniques for influence that fall along a continuum between compliance and commitment. The ten techniques described below seek different degrees of compliance or commitment ranging from pressure at the compliance end to relations building at the commitment end.

7-8. **Pressure** is applied when leaders use explicit demands to achieve compliance, such as establishing task completion deadlines with negative consequences imposed for unmet completion. Indirect pressure includes persistent reminders of the request and frequent checking. This technique should be used infrequently since it tends to trigger resentment from followers, especially if the leader-exerted pressure becomes too severe. When followers perceive that pressures are not mission related but originate from their leader's attempt to please superiors for personal recognition, resentment can quickly undermine an organization's morale, cohesion, and quality of performance. Pressure is a good choice when the stakes are high, time is short, and previous attempts at achieving commitment have not been successful.

7-9. **Legitimate requests** occur when leaders refer to their source of authority to establish the basis for a request. In the military, certain jobs must be done regardless of circumstances when subordinate leaders receive legitimate orders from higher headquarters. Reference to one's position suggests to those who are being influenced that there is the potential for official action if the request is not completed.

7-10. **Exchange** is an influence technique that leaders use when they make an offer to provide some desired item or action in trade for compliance with a request. The exchange technique requires that the leaders control certain resources or rewards that are valued by those being influenced. A four-day pass as reward for excelling during a maintenance inspection is an example of an exchange influence technique.

7-11. **Personal appeals** occur when the leader asks the follower to comply with a request based on friendship or loyalty. This might often be useful in a difficult situation when mutual trust is the key to success. The leader appeals to the follower by highlighting the subordinate leader's special talents and professional trust to strengthen him prior to taking on a tough mission. An S3 might ask a staff officer to brief at an important commander's conference if the S3 knows the staff officer will do the best job and convey the commander's intent.

7-12. **Collaboration** occurs when the leader cooperates in providing assistance or resources to carry out a directive or request. The leader makes the choice more attractive by being prepared to step in and resolve any problems. A major planning effort prior to a deployment for humanitarian assistance would require possible collaboration with joint, interagency, or multinational agencies.

7-13. **Rational persuasion** requires the leader to provide evidence, logical arguments, or explanations showing how a request is relevant to the goal. This is often the first approach to gaining compliance or commitment from followers and is likely to be effective if the leader is recognized as an expert in the specialty area in which the influence occurs. Leaders often draw from their own experience to give reasons that some task can be readily accomplished because the leader has tried it and done it.

7-14. **Apprising** happens when the leader explains why a request will benefit a follower, such as giving them greater satisfaction in their work or performing a task a certain way that will save half the time. In contrast to the exchange technique, the benefits are out of the control of the leader. A commander may use the apprising technique to inform a newly assigned noncommissioned officer that serving in an operational staff position, prior to serving as a platoon sergeant, could provide him with invaluable experience. The commander points out that the additional knowledge may help the NCO achieve higher performance than his peers and possibly lead to an accelerated promotion to first sergeant.

7-15. **Inspiration** occurs when the leader fires up enthusiasm for a request by arousing strong emotions to build conviction. A leader may stress to a fellow officer that without help, the safety of the team may be at risk. By appropriately stressing the results of stronger commitment, a unit leader can inspire followers to surpass minimal standards and reach elite performance status.

7-16. **Participation** occurs when the leader asks a follower to take part in planning how to address a problem or meet an objective. Active participation leads to an increased sense of worth and recognition. It provides value to the effort and builds commitment to execute the commitment. Invitation to get involved is critical when senior leaders try to institutionalize a vision for long-term change. By involving key leaders of all levels during the planning phases, senior leaders ensure that their followers take stock in the vision. These subordinates will later be able to pursue critical intermediate and long-term objectives, even after senior leaders have moved on.

7-17. **Relationship building** is a technique in which leaders build positive rapport and a relationship of mutual trust, making followers more willing to support requests. Examples include, showing personal interest in a follower's well-being, offering praise, and understanding a follower's perspective. This technique is best used over time. It is unrealistic to expect it can be applied hastily when it has not been previously used. With time, this approach can be a consistently effective way to gain commitment from followers.

Putting Influence Techniques to Work

7-18. To succeed and create true commitment, influencing techniques should be perceived as authentic and sincere. Positive influence comes from leaders who do what is right for the Army, the mission, the team, and each individual Soldier. Negative influence—real and perceived—emanates from leaders who primarily focus on personal gain and lack self-awareness. Even honorable intentions, if wrongly perceived by followers as self-serving, will yield mere compliance. False perception may trigger unintended side effects such as resentment of the leader and the deterioration of unit cohesion.

7-19. The critical nature of the mission also determines which influence technique or combination of techniques is appropriate. When a situation is urgent and greater risk is involved, eliciting follower compliance may be desirable. Direct-level leaders often use compliance techniques to coordinate team activities in an expedient manner. In comparison, organizational leaders typically pursue a longer-term focus and use indirect influence to build strong commitment.

7-20. When influencing their followers, Army leaders should consider that—

- The objectives for the use of influence should be in line with the Army Values, ethics, the Uniform Code of Military Justice, the Warrior Ethos, and the Civilian Creed.
- Various influence techniques can be used to obtain compliance and commitment.
- Compliance-seeking influence focuses on meeting and accounting for specific task demands.
- Commitment-encouraging influence emphasizes empowerment and long-lasting trust.

Providing Purpose, Motivation, and Inspiration

7-21. Leaders influence others to achieve some purpose. To be successful at exerting influence Army leaders have an end or goal in mind. Sometimes the goal will be very specific, like reducing the number of training accidents by one-half over a period of six months. Many goals are less distinct and measurable than this example, but are still valid and meaningful. A leader may decide that unit morale needs to be improved and may set that as a goal for others to join to support.

7-22. Purpose provides what the leader wants done, while motivation and inspiration provide the energizing force to see that the purpose is addressed and has the strength to mobilize and sustain effort to get the job done. Motivation and inspiration address the needs of the individual and team. Indirect needs—like job satisfaction, sense of accomplishment, group belonging, and pride—typically have broader reaching effects than formal rewards and punishment, like promotions or nonjudicial actions.

7-23. Besides purpose and motivation, leader influence also consists of direction. Direction deals with how a goal, task, or mission is to be achieved. Subordinates do not need to receive guidance on the details of execution in all situations. The skilled leader will know when to provide detailed guidance and when to focus only on purpose, motivation, or inspiration.

7-24. Mission command conveys purpose without providing excessive, detailed direction. Mission command is the conduct of military operations through decentralized execution based on mission orders for effective mission accomplishment. Successful mission command rests on four elements:

- Commander's intent.
- Subordinates' initiative.
- Mission orders.
- Resource allocation.

7-25. Mission command is a basis for Army planning (as described in FM 5-0) and is thoroughly explained in FM 6-0.

Providing Purpose

7-26. Leaders in command positions use commander's intent to convey purpose. The *commander's intent* is a clear, concise statement of what the force must do and the conditions the force must meet to succeed with respect to the enemy, terrain, and desired end state (FM 3-0). When leading in other than command positions or in a nontactical application, leaders also establish tasks and the conditions for successful accomplishment. For leader situations other than command and for Army civilian leaders, enemy and terrain may be substituted by factors such as goals or organizational obstacles. Leaders communicate purpose with implied or explicit instructions so that others may exercise initiative while maintaining focus. This is important for situations when unanticipated opportunities arise or the original solution no longer applies. While direct and organizational level leaders provide purpose or intent, strategic leaders usually provide long-term vision or conceptual models.

Motivating and Inspiring

7-27. Motivation is the reason for doing something or the level of enthusiasm for doing it. Motivation comes from an inner desire to put effort into meeting a need. People have a range of needs. They include basics, such as survival and security and advanced needs, such as belonging and a sense of accomplishment. Awareness of one's own needs is most acute when needs go unfulfilled.

7-28. Army leaders use the knowledge of what motivates others to influence those they lead. Knowing one's Soldiers and others who may be influenced, gives leaders insight into guiding the team to higher levels of performance. Understanding how motivation works provides insight into why people may take action and how strongly they are driven to act.

7-29. While it is difficult to know others' needs, it helps to consider three parts that define motivation:

- Arousal: A need or desire for something that is unfulfilled or below expectations.
- Direction: Goals or other guides that direct the course of effort and behavior.
- Intensity: The amount of effort that is applied to meet a need or reach a goal.

7-30. The arousal, direction, and intensity of motivation produce at least four things that contribute directly to effective task performance. Motivation focuses **attention** on issues, goals, task procedures, or other aspects of what needs to be done. Motivation produces **effort** that dictates how hard one tries. Motivation generates **persistence** in terms of how long one tries. The fourth product of motivation is **task strategies** that define how a task is performed—the knowledge and skills used to reach a particular goal. Knowing better ways to perform a task can improve performance and lead to success in reaching a desired goal.

7-31. Motivation is based on the individual and the situation. Individuals contribute job knowledge and ability, personality and mood, and beliefs and values. The situation is the physical environment, task procedures and standards, rewards and reinforcements, social norms, and organization climate and culture. Leaders can improve individual motivation by influencing the individual and the situation. The influence techniques operate on different parts of motivation.

7-32. Self-efficacy is the confidence in one's ability to succeed at a task or reach a goal. Leaders can improve others' motivation by enhancing their self-efficacy by developing necessary knowledge and skills. Certain knowledge and skills may contribute to working smarter and just working harder or longer. An example is learning a more effective way to perform a task without reducing the quality of work.

7-33. Emotional inspiration is another way that a leader can enhance motivation. Providing an inspirational vision of future goals can increase the inner desire of a subordinate to achieve that vision. Leaders can inspire through the images when speaking. Inspirational images energize the team to go beyond satisfying individual interests and exceed expectations. Combat and life-threatening situations cause enough arousal as a natural response that leaders in these situations do not need to energize. Instead, they need to moderate too much arousal by providing a steady and calming influence and focus. Creating the right level of emotional arousal takes a careful balancing act. Training under severe and stressful conditions allows individuals the chance to experience different levels of arousal.

7-34. Leaders can encourage subordinates to set goals on their own and to set goals together. When goals are accepted they help to focus attention and action, increase the effort that is expended and persistence even in the face of failure, and develop strategies to help in goal accomplishment.

7-35. Positive reinforcement in the form of incentives (for example, monetary rewards or time off) as well as internal rewards (for example, praise and recognition) can enhance motivation. Punishment can be used when there is an immediate need to discontinue dangerous or otherwise undesirable behavior. Punishment can also send a clear message to others in the unit about behavioral expectations and the consequences of violating those expectations. In this way, a leader can shape the social norms of a unit. One caution is that punishment should be used sparingly and only in extreme cases because it can lead to resentment.

7-36. Effective leaders leverage the values and shared goals of those within their sphere of influence in order to motivate others. Leaders encourage others to reflect on their commitments such as the shared goals in this unit. Additionally, there are often shared values within an organization that form the basis of individual commitments (for example, personal courage, honor, and loyalty). Letting others know how a particular task is related to a larger mission, objective, or goal is often an effective motivational technique.

7-37. Individuals can be motivated by the duties they perform. Generally, if someone enjoys performing a task and is internally motivated, the simple acknowledgment of a job well done may be enough to sustain performance. No other rewards or incentives are necessary to motivate continued work on the task. In this case, task enjoyment provides the internal reward that motivates a Soldier to complete a task.

7-38. People often want to be given the opportunity to be responsible for their own work and to be creative—they want to be empowered. Empower subordinates by training them to do a job and providing them with necessary task strategies; give them the necessary resources, authority and clear intent; and then step aside to let them accomplish the mission. Empowering subordinates is a forceful statement of trust and one of the best ways of developing them as leaders. It is important to point out that being empowered also implies accepting the responsibility for the freedom to act and create.

7-39. Effective motivation is achieved when the team or organization wants to succeed. Motivation involves using words and examples to inspire subordinates to accomplish the mission. It grows from people's confidence in themselves, their unit, and their leaders. That confidence develops through tough and realistic training as well as consistent and fair leadership. Motivation also springs from the person's faith in the organization's larger missions, a sense of being a part of the bigger picture.

Building and Sustaining Morale

There is a soul to an army as well as to the individual man, and no general can accomplish the full work of his army unless he commands the soul of his men as well as their bodies and legs.

General William T. Sherman
Letter to General Ulysses S. Grant

7-40. Military historians describing great armies often focus on weapons and equipment, training, and the National cause. They may mention numbers or other factors that can be analyzed, measured, and compared. Many historians also place great emphasis on one critical factor that cannot be easily measured: the emotional element called morale.

7-41. Morale is the human dimension's most important intangible element. It is a measure of how people feel about themselves, their team, and their leaders. High morale comes from good leadership, shared effort, and mutual respect. An emotional bond springs from the Warrior Ethos, common values like loyalty, and a belief that the Army will care for Soldiers' families. High morale results in a cohesive team striving to achieve common goals. Competent leaders know that morale—the essential human element—holds the team together and keeps it going in the face of the terrifying and dispiriting things that occur in war.

7-42. Captain Audie Murphy, Medal of Honor recipient and most decorated Soldier of World War II, puts morale in the following simple words:

You have a comradeship … a rapport that you'll never have again, not in our society, anyway. I suppose it comes from having nothing to gain except the end of the war. There's no competitiveness, no money values. You trust the man on your left and on your right with your life, while, as a civilian, you might not trust either one of them with ten cents.

7-43. One unit that represented the Army's expectations of enduring high morale was Easy Company, 506th Parachute Infantry Regiment, 101st Airborne Division. Major Richard Winters commanded the company from the Normandy Invasion to the defeat of Germany in 1945. In a recorded interview, he stressed that good morale results from mutual respect between leaders and followers—the leader living with his men and knowing them. He emphasized that good leaders must be prepared to give to the people they lead—in every way. They should never take from the people they lead.

7-44. Without a doubt, Easy Company's high morale grew from a strong mutual trust in most of their leaders and the many friendships forged during training and actual combat. The unit knew its commander would go to bat for them to maintain a balance between combat readiness and the need for recovery and relaxation. Easy Company's example shows that commanders can build morale by carefully balancing hard work and sacrifice in combat with appropriate recognition and rewards. The rewards can be simple things, such as a good night's sleep away from the front, warm meals, phone calls home, and movies. Rewards can also include extended leave periods and morale, welfare, and recreation sponsored trips.

7-45. Leaders can furthermore boost morale in the face of extreme danger by providing their Soldiers the force protection means and support for successful operations. Units with high morale are usually more effective in combat and deal with hardships and losses better. It does not come as a surprise that these units conduct reunions and maintain close friendships for decades after they have served together in combat. A message scribbled by an Army aviator in distress during the Somalia operations in 1993 reminds us that exceptional morale is always present in our Army's Soldiers and well-led units. When Chief Warrant Officer Mike Durant was injured and held captive by Somali guerillas in October 1993, he wrote his wife:

NSDQ = Night Stalkers Don't Quit!

Motto of the 160th Special Operations
Aviation Regiment, "The Night Stalkers"

ENFORCING STANDARDS

7-46. To lead others and gauge if a job has been done correctly, the Army has established standards for military activities. Standards are formal, detailed instructions that can be described, measured, and achieved. They provide a mark for performance to assess how a specific task has been executed. To use standards effectively, leaders know, communicate, and enforce high but realistic standards. Good leaders explain the standards that apply to their organizations, but give subordinates the authority to enforce them.

7-47. When enforcing standards for unit activities, leaders must remain aware that not everything can be a number one priority. Striving for excellence in every area, regardless of how trivial, would work an organization too hard. Leaders must prioritize the tasks without allowing other tasks to drop below established standards. True professionals make sure the standard fits the task's importance.

7-48. A leader's ultimate goal is to train the organization to the standards that ensure success in its wartime mission. The leader's daily work includes setting the intermediate goals to prepare the organization to reach the standards. To be successful at this, leaders use the Army training management cycle. The training management process is used to set appropriate training goals and to plan, resource, execute, and evaluate training accordingly (see FM 7-0 for more detail).

Performing Checks and Inspections

7-49. Proper supervision is essential to ensuring mission accomplishment to standard. It is an integral part of caring for Soldiers. The better they know their unit and subordinates, the more they can strike a balance for finding the details. Training subordinates for independent action is vital. To foster independence and initiative, direct leaders give instructions and clear mission intent. Then they allow subordinates to get the work done without constantly looking over their shoulders.

7-50. Accomplishing the unit's real-world mission is critically important. This requires that units and individuals are fully prepared. It is why leaders check things—conducting pre-operation checks and formal inspections (FM 6-0). Thorough inspections ensure that Soldiers, units, and systems are as fully capable and ready to execute the mission as time and resources permit.

7-51. Focused checking minimizes the chance of neglect or mistakes that may derail a mission or cause unnecessary casualties. Checking also gives leaders a chance to see and recognize subordinates who are doing things right or to make on-the-spot corrections when necessary. For example, a platoon sergeant delegates to the platoon's squad leaders the authority to get their squads ready for a tactical road march. The platoon sergeant oversees the activity but does not personally intervene unless errors, sloppy work, or lapses in performance occur. The platoon sergeant is mainly present to answer questions or resolve problems that the squad leaders cannot handle. This type of supervision ensures that the squads are prepared to meet standards, while giving the squad leaders the authority and confidence to do their job.

Instilling Discipline

7-52. Leaders who consistently enforce standards are simultaneously instilling discipline that will pay-off in critical situations. Disciplined people take the right action, even if they do not feel like it. True discipline demands habitual and reasoned obedience, an obedience that preserves initiative and works, even when the leader is not around or when chaos and uncertainty abound.

7-53. Discipline does not mean barking orders and demanding instant responses. A good leader gradually instills discipline by training to standard, using rewards and punishment judiciously, instilling confidence, building trust among team members, and ensuring that Soldiers and civilians have necessary technical and tactical expertise. Confidence, trust, and team effort are crucial for success in operational settings.

7-54. Individual and collective discipline generally carries the day when organizations are faced with complex and dangerous situations. It usually begins with the resilience, competence, and discipline of one individual who recognizes the need to inspire others to follow an example, turning a negative situation into success. One such event during Operation Iraqi Freedom showed how discipline during training could make the difference during wartime.

One Man Can Make a Difference

When SFC Paul Smith started his day at the Baghdad Airport on 4 April 2003, he was focused on building a holding pen for enemy prisoners. Before the day was over, he had given his life and saved as many as 100 others in the process.

SFC Smith was a combat engineer assigned to Bravo Company, 11th Engineer Battalion in support of Task Force 2-7 Infantry. Smith, whose call sign was "Sapper 7," was well liked by his Soldiers. He was a taskmaster and his experience in Desert Storm taught him to train tirelessly and to standard. He was the acting platoon leader when 50-100 of Saddam's well-trained Republican Guard attacked him and his men.

Three of his Soldiers were seriously wounded and Smith helped evacuate them to a nearby aid station that was also threatened by the attackers. He organized a hasty defense. He told one of his Soldiers "we are in a world of hurt."

Selflessly, Smith took over a .50 caliber machine gun in an exposed position. He fired over 300 rounds at the enemy before his gun fell silent. SFC Smith was the only member of his unit to die that day. For his discipline and courage under fire, he received the first Medal of Honor awarded during Operation Iraqi Freedom.

7-55. Soldiers have overcome treacherous ambush situations throughout history. Like Sergeant First Class Paul Smith, all possessed the unique ability to persevere in adversity. That ability is deeply rooted in confidence in themselves, their friends, their leaders, their equipment, and their training. Most importantly, Soldiers endure because they have discipline and are resilient.

BALANCING MISSION AND WELFARE OF SOLDIERS

Leading and caring are essential to readiness and excellence. . . .

General John A. Wickham, Jr.
Chief of Staff, Army (1983-1987)

7-56. Consideration of the needs of Soldiers and civilians is a basic function of all Army leaders. Having genuine concern for the well-being of followers goes hand-in-hand with motivation, inspiration, and influence. Soldiers and civilians will be more willing to go the extra mile for leaders who they know look out for them. Sending Soldiers or civilians in harm's way to accomplish the mission seems to contradict all the emphasis on taking care of people. How can a leader truly care for comrades and send them on missions that might get them killed? Similarly, when asking junior officers and NCOs to define what leaders do, the most common response is, "Take care of Soldiers."

7-57. Taking care of Soldiers entails creating a disciplined environment where they can learn and grow. It means holding them to high standards when training and preparing them to do their jobs so they can succeed in peace and win in war. Taking care of Soldiers, treating them fairly, refusing to cut corners, sharing hardships, and setting a personal example are crucial.

7-58. Taking care of Soldiers also means demanding that Soldiers do their duty—even at risk to their lives. Preparing Soldiers for the brutal realities of actual combat is a direct leader's most important duty. It does not mean coddling or making training easy or comfortable. Training neglect of that kind can get Soldiers killed. Training must be rigorous and simulate combat as much as possible, while keeping safety in mind. Leaders use risk management to ensure safety standards are appropriate. During wartime operations, unit leaders must also recognize the need to provide Soldiers with reasonable comforts to bolster morale and maintain long-term combat effectiveness. Comfort always takes second seat to the mission.

7-59. Taking care of others means finding out a Soldier's personal state on a particular day or their attitude about a particular task. The three attributes of a leader—character, presence, and intellectual capacity—can be applied as a leader's mental checklist to check on the welfare and readiness of Soldiers and civilians alike. It is up to the leader to provide the encouragement to push through to task completion or, when relief is required, to prevent unacceptable risk or harm and find other means to accomplish the task.

7-60. Many leaders connect at a personal level with their followers so they will be able to anticipate and understand the individual's circumstances and needs. As discussed previously in the chapter, building relationships is one way to gain influence and commitment from followers. Knowing others is the basis that many successful leaders use to treat personnel well. It includes everything from making sure a Soldier has time for an annual dental exam, to finding out about a person's preferred hobbies and pastimes. Leaders should provide an adequate family support and readiness network that assures Soldiers' families will be taken care of, whether the Soldier is working at home station or deployed.

EXTENDS INFLUENCE BEYOND THE CHAIN OF COMMAND

7-61. While Army leaders traditionally exert influence within their unit and its established chain of command, multiskilled leaders must also be capable of extending influence to others beyond the chain of command. Extending influence is the second leader competency. In today's politically and culturally charged operational environments, even direct leaders may work closely with joint, interagency, and multinational forces, the media, local civilians, political leaders, police forces, and nongovernmental agencies. Extending influence requires special awareness about the differences in how influence works.

7-62. When extending influence beyond the traditional chain, leaders often have to influence without authority designated or implied by rank or position. Civilian and military leaders often find themselves in situations where they must build informal teams to accomplish organizational tasks.

7-63. A unique aspect of extending influence is that those who are targets of influence outside the chain may not even recognize or willingly accept the authority that an Army leader has. Often informal teams must be created in situations where there are no official chains of authority. In some cases, it may require leaders to establish their credentials and capability for leading others. At other times, leaders may need to interact as a persuasive force but not from an obvious position and attitude of power.

7-64. The key element of extending influence and building teams is the creation of a common vision among prospective team members. At times leaders may need to interact with others as a persuasive influence but not from an obvious position and attitude of power.

7-65. Leading without authority requires adaptation to the environment and cultural sensitivities of the given situation. Leaders require cultural knowledge to understand different social customs and belief systems and to address issues in those contexts. When conducting peace operations, for example, even small unit leaders and civilian negotiators must understand that their interaction with locals and their leaders can have dramatic impacts on the overall theater strategy. The manner in which a unit conducts house-to-house searches for insurgents can influence the local population's acceptance of authority, or become a recruiting incentive for the insurgency.

7-66. Extending influence includes the following competency subsets:

- Building trust outside lines of military command authority.
- Understanding the sphere, means, and limits of influence.
- Negotiating, consensus building, and conflict resolution.

Building Trust Outside Lines Of Authority

7-67. Forming effective, cohesive teams is often the first challenge of a leader working outside a traditional command structure. These teams usually have to be formed from disparate groups who are unfamiliar with military and Army customs and culture. Without some measure of trust, nothing will work as well. To establish trust, the leader will have to identify areas of common interests and goals. Trust between two people or two groups is based largely on being able to anticipate what the others understand and how they will respond in various situations. Keeping others informed also builds trust. Cementing and sustaining trust depends on following through on commitments.

7-68. Successful teams develop an infectious winner's attitude. Problems are challenges rather than obstacles. Cohesive teams accomplish missions much more efficiently than a loose group of individuals. While developing seamless teams is ideal, sometimes it will not be practical to bring disparate groups together.

7-69. Building alliances is similar to building teams; the difference being that in alliances the groups maintain greater independence. Trust is a common ingredient in effective alliances. Alliances are groomed over time by establishing contact with others, growing friendships, and identifying common interests.

7-70. Whether operating in focused teams or in looser alliances, training and working together builds collective competence and mutual trust. A mutual trust relationship will ultimately permeate the entire organization, embracing every single member, regardless of gender, race, social origin, religion or if permanently assigned or temporarily attached.

7-71. The requirements for building trust and cohesion are valid for relationships extending beyond the organization and the chain of command. They apply when working with task-organized organizations; joint, interagency, and multinational forces; and noncombatants. If a special operations team promises critical air support and medical supplies to indigenous multinational forces for an upcoming operation, the personal reputation of the leader, and trust in the United States as a respected, supportive nation, can be at stake.

Understanding Sphere, Means, and Limits of Influence

7-72. When operating with an established command structure and common procedures, the provisions and limits of roles and responsibilities are readily apparent. When leading outside an established organization, assessing the parties involved becomes another part of the operation. Identifying who is who, what role they have, over whom they have authority or influence, and how they are likely to respond to the Army leader's influence are all important considerations. Sometimes this is viewed as understanding the limits to the Army's or the leader's influence.

7-73. Spanning the boundaries of disparate groups or organizations is a task that requires special attention. The key to influence outside the chain of command is to learn about the people and organizations. By understanding their interests and desires, the leader will know what influence techniques are most likely to work. Leaders can learn some of the art of dealing with disparate interests from business operations that deal with coordinating opposing parties with different interests.

Negotiating, Building Consensus, and Resolving Conflicts

7-74. While operating outside the chain of command, leaders often have to resolve conflicts between Army interests and local populations or others. Conflict resolution identifies differences and similarities among the stances of the various groups. Differences are further analyzed to understand what is behind the difference. Proposals are made for re-interpreting the differences or negotiating compromise to reach common understanding or shared goals. Trust, understanding, and knowing the right influence technique for the situation are the determining factors in negotiating, consensus building, and conflict resolution.

LEADS BY EXAMPLE

DISPLAYING CHARACTER

7-75. Leaders set an example whether they know it or not. Countless times leaders operate on instinct that has grown from what they have seen in the past. What leaders see others do sets the stage for what they may do in the future. A leader of sound character will exhibit that character at all times. Modeling these attributes of character defines the leaders to the people with whom they interact. A leader of character does not have to worry about being seen at the wrong moment doing the wrong thing.

7-76. Living by the Army Values and the Warrior Ethos best displays character and leading by example. It means putting the organization and subordinates above personal self-interest, career, and comfort. For the Army leader, it requires putting the lives of others above a personal desire for self-preservation.

Leading with Confidence in Adverse Conditions

7-77. A leader who projects confidence is an inspiration to followers. Soldiers will follow leaders who are comfortable with their own abilities and will question the leader who shows doubt.

7-78. Displaying confidence and composure when things are not going well can be a challenge for anyone, but is important for the leader to lead others through a grave situation. Confidence is a key component of leader presence. A leader who shows hesitation in the face of setbacks can trigger a chain reaction among others. A leader who is over-confident in difficult situations may lack the proper degree of care or concern.

7-79. Leading with confidence requires a heightened self-awareness and ability to master emotions. Developing the ability to remain confident no matter what the situation involves—

- Having prior opportunities to experience reactions to severe situations.
- Maintaining a positive outlook when a situation becomes confusing or changes.
- Remaining decisive after mistakes have been discovered.
- Encouraging others when they show signs of weakness.

Displaying Moral Courage

7-80. Projecting confidence in combat and other situations requires physical and moral courage. While physical courage allows infantrymen to defend their ground, even when the enemy has broken the line of defense and ammunition runs critically short, moral courage empowers leaders to stand firm on values, principles, and convictions in the same situation. Leaders who take full responsibility for their decisions and actions display moral courage. Morally courageous leaders are willing to critically look inside themselves, consider new ideas, and change what caused failure.

7-81. Moral courage in day-to-day peacetime operations is as important as momentary physical courage in combat. Consider a civilian test board director who has the responsibility to determine whether a new piece of military equipment performs to the established specifications. Knowing that a failed test may cause the possibility of personal pressure and command resistance from the program management office, a morally courageous tester will be prepared to endure that pressure and remain objective and fair in test procedures and conclusions. Moral courage is fundamental to living the Army Values of integrity and honor, whether a civilian or military team member.

Demonstrating Competence

7-82. It does not take long for followers to become suspicious of a leader who acts confident but does not have the competence to back it up. Having the appropriate levels of domain knowledge is vital to prepare competent leaders who can in turn display confidence through their attitudes, actions, and words.

7-83. When examining the majority of small unit military operations, many often were uncertain until competent and confident leaders made the difference. At the right time, the competent leaders apply the decisive characteristics to influence the tactical or operational situation. Their personal presence and indirect influences help mobilize the will and morale in their people to achieve final victory.

7-84. Leading by example demands that leaders stay aware of how their guidance and plans are executed. Direct and organizational leaders cannot remain in safe, dry headquarters, designing complex plans without examining what their Soldiers and civilians are experiencing. They must have courage to get out to where the action is, whether the battlefield or the shop floor. Good leaders connect with their followers by sharing hardships and communicating openly to clearly see and feel what goes on from a subordinate's perspective.

7-85. Military leaders at all levels must remember that graphics on a map symbolize human Soldiers, often fighting at very close range. To verify that a plan can succeed, true warrior leaders lead from the front and share the experiences of their Soldiers. Seeing and feeling the plan transform into action empowers the leader to better assess the situation and influence the execution by their immediate presence. Leaders who stay at a safe distance from the front risk destroying their Soldiers' trust and confidence. Similar concerns apply for civilian leaders when operating under difficult conditions, such as 24/7 maintenance operations or dangerous supply missions in support of deployed military forces. Just like their counterparts in uniform, they must ask themselves: Would I readily do what I'm asking my workers to do?

7-86. General Patton made it clear that leading from the front and making plans with a clear understanding of the frontline situation were keys to success. In his General Orders to the 3rd Army of 6 March 1944, he stipulated:

> *The Commanding General or his Chief of Staff (never both at once) and one member of each of the General Staff sections, the Signal, Medical, Ordnance, Engineer, and Quartermaster sections, should visit the front daily. To save duplication, the Chief of Staff will designate the sector each is to visit.*
>
> *The function of these Staff officers is to observe, not to meddle. In addition to their own specialty, they must observe and report anything of military importance. ...Remember, too, that your primary mission as a leader is to see with your own eyes and be seen by your troops while engaged in personal reconnaissance.*

COMMUNICATES

7-87. Competent leadership that gets results depends on good communication. Although communication is usually viewed as a process of providing information, communication as a competency must ensure that there is more than the simple transmission of information. Communication needs to achieve a new understanding. Communication must create new or better awareness. Communicating critical information in a clear fashion is an important skill to reach a shared understanding of issues and solutions. It is conveying thoughts, presenting recommendations, bridging cultural sensitivities and reaching consensus. Leaders cannot lead, supervise, build teams, counsel, coach, or mentor without the ability to communicate clearly.

Listening Actively

7-88. An important form of two-way communication to reach a shared understanding is active listening. Although the most important purpose of listening is to comprehend the sender's thoughts, listeners should provide an occasional indication to the speaker that they are still attentive. Active listening involves avoiding interruption and keeping mental or written notes of important points or items for clarification.

Good listeners will be aware of the content of the message, but also the urgency and emotion of how it is spoken.

7-89. It is critical to remain aware of barriers to listening. Do not formulate a response while it prevents hearing what the other person is saying. Do not allow distraction by anger, disagreement with the speaker, or other things to impede. These barriers prevent hearing and absorbing what is said.

STATING GOALS FOR ACTION

7-90. The basis for expressing clear goals for action resides in the leader's vision and how well that vision is explained. Before stating goals, objectives, and required tasks for the team, unit, or organization, it is important for the leader to visualize a desired end state. Once the goals are clear, leaders communicate them in a way that motivates them to understand the message and to accept and act on the message.

7-91. Speaking to engage listeners can improve by being aware of what styles of communication energize the leader when the leader is the listener. The speaker should be open to cues that listeners give and adapt to ensure that his message is received. The speaker needs to be alert to recognize and resolve misunderstandings. Since success or failure of any communication is the leader's responsibility, it is important to ensure the message has been received. Leaders can use backbriefs or ask a few focused questions to do so.

ENSURING SHARED UNDERSTANDING

7-92. Competent leaders know themselves, the mission, and the message. They owe it to their organization and their subordinates to share information that directly applies to their duties. They should also provide information that provides context for what needs to be done. Generous sharing of information also provides information that may be useful in the future.

7-93. Leaders keep their organizations informed because it builds trust. Shared information helps relieve stress and control rumors. Timely information exchange allows team members to determine what needs to be done to accomplish the mission and adjust to changing circumstances. Informing subordinates of a decision, and the overall reasons for it, shows they are appreciated members of the team and conveys that support and input are needed. Good information flow also ensures the next leader in the chain can be sufficiently prepared to take over, if required. Subordinates must clearly understand the leader's vision. In a tactical setting, all leaders must fully understand their commanders' intent two levels up.

7-94. Leaders use a variety of means to share information: face-to-face talks, written and verbal orders, estimates and plans, published memos, electronic mail, websites, and newsletters. When communicating to share information, the leader must acknowledge two critical factors:

- A leader is responsible for making sure the team understands the message.
- A leader must ensure that communication is not limited to the traditional chain of command but often includes lateral and vertical support networks.

7-95. When checking the information flow for shared understanding, a team leader should carefully listen to what supervisors, platoon sergeants, platoon leaders, and company commanders say. A platoon sergeant who usually passes the message through squad leaders or section chiefs should watch and listen to the troops to verify that the critical information makes it to where it will ultimately be translated into action.

7-96. Communicating also flows from bottom to top. Leaders find out what their people are thinking, saying, and doing by listening. Good leaders keep a finger on the pulse of their organizations by getting out to coach, to listen, and to clarify. They then pass relevant observations to their superiors who can assist with planning and decision making.

7-97. Often, leaders communicate more effectively with informal networks than directly with superiors. Sometimes that produces the desired results but can lead to misunderstandings and false judgments. To run an effective organization and achieve mission accomplishment without excessive conflict, leaders must figure out how to reach their superiors when necessary and to build a relationship of mutual trust. First, leaders must assess how the boss communicates and how information is received. Some use direct and personal contact while others may be more comfortable with weekly meetings, electronic mail, or memoranda. Knowing the boss's intent, priorities, and thought processes enhance organizational effectiveness and success. A leader who communicates well with superiors minimizes friction and improves the overall organizational climate.

7-98. To prepare organizations for inevitable communication challenges, leaders create training situations where they are forced to act with minimum guidance or only the commander's intent. Leaders provide formal or informal feedback to highlight the things subordinates did well, what they could have done better, and what they should do differently next time to improve information sharing and processing.

7-99. Open communication does more than share information. It shows that leaders care about those they work with. Competent and confident leaders encourage open dialogue, listen actively to all perspectives, and ensure that others can voice forthright and honest opinions, without fear of negative consequences.

Chapter 8
Developing

8-1. Good leaders strive to leave an organization better than they found it and expect other leaders throughout the Army do the same. Leaders can create a positive organizational climate, prepare themselves to do well in their own duties, and help others to perform well. Good leaders look ahead and prepare talented Soldiers and civilians to assume positions with greater leadership responsibility in their own organization and in future assignments. They also work on their own development to prepare for new challenges.

8-2. To have future focus and maintain balance in the present, Army leaders set priorities and weigh competing demands. They carefully steer their organizations' efforts to address short- and long-term goals, while continuing to meet requirements that could contribute directly to achieving those goals. Accounting for the other demands that vie for an organization's time and resources, a leader's job becomes quite difficult. Guidance from higher headquarters may help, but leaders have to make the tough calls to keep a healthy balance.

8-3. Developing people and the organization with a long-term perspective requires the following:

- The leader must create a positive environment that fosters teamwork, promotes cohesion, and encourages initiative and acceptance of responsibility. A leader should also maintain a healthy balance between caring for people and focusing on the mission.
- The leader must seek self-improvement. To master the profession at every level, a leader must make a full commitment to lifelong learning. Self-improvement leads to new skills necessary to adapt to changes in the leadership environment. Self-improvement requires self-awareness.
- The leader must invest adequate time and effort to develop individual subordinates and build effective teams. Success demands a fine balance of teaching, counseling, coaching, and mentoring.

CREATES A POSITIVE ENVIRONMENT

8-4. Climate and culture describe the environment in which a leader leads. Culture refers to the environment of the Army as an institution and of major elements or communities within it. While strategic leaders maintain the Army's institutional culture, climate refers to the environment of units and organizations, primarily shaped by organizational and direct leaders.

8-5. Taking care of people and maximizing their performance is influenced by how well the leader shapes the organization's climate. Climate is how members feel about the organization and comes from shared perceptions and attitudes about the unit's daily functioning. These things have a great impact on their motivation and the trust they feel for their team and their leaders. Climate is generally a short-term experience, depending on a network of the personalities in a small organization. The organization's climate changes as people come and go. When a Soldier says, "My last platoon sergeant was pretty good, but this new one is great," the Soldier is pinpointing one of the many elements that affect an organization's climate.

8-6. Culture is a longer lasting and more complex set of shared expectations than climate. While climate is a reflection about how people think and feel about their organization right now, culture consists of the shared attitudes, values, goals, and practices that characterize the larger institution over time. It is deeply rooted in long-held beliefs, customs, and practices. Leaders must establish a climate consistent with the culture of the enduring institution. They also use the culture to let their people know they are part of something bigger than just themselves, that they have responsibilities not only to the people around them but also to those who have gone before and those who will come after.

8-7. Soldiers draw strength from knowing they are part of a long-standing tradition. Most meaningful traditions have their roots in the institution's culture. Many of the Army's everyday customs and traditions exist to remind Soldiers they are the latest addition to a long line of Soldiers. Army culture and traditions connect Soldiers to the past and to the future. The uniforms, the music played during official ceremonies, the way Soldiers salute, military titles, the organization's history, and the Army Values all are reminders of a place in history. This sense of belonging lives in many veterans long after they have left the service. For most, service to the Nation remains the single most significant experience of their lives.

8-8. Soldiers join the Army to become part of a values and tradition based culture. While the Army Values help deepen existing personal values, such as family bonds, work ethic, and integrity, it is tradition that ties Soldiers and their families into military culture. Unit history is an important factor for that bonding, since Soldiers want to belong to organizations with distinguished service records. Unit names, such as the Big Red One, Old Ironsides, All Americans, and Spearhead carry an extensive history. To sustain tradition, leaders must teach Soldiers the history that surrounds unit crests, military greetings, awards, decorations, and badges. Through leading by example, teaching, and upholding traditions, leaders ensure that the Army's culture becomes an integral part of every member of the Army team and adds purpose to their lives.

SETTING THE CONDITIONS FOR POSITIVE CLIMATE

8-9. Climate and culture are the context in which leaders and followers interact. Each element has an effect on the other. Research in military, government, and business organizations shows that a positive environment leads to workers who feel better about themselves, have stronger commitments, and produce better work. If leaders set the tone for a positive climate, others will respond in kind.

8-10. Good leaders are concerned with establishing a climate that can be characterized as fair, inclusive, and ethical. Fairness means that treatment is equitable and no one gets preferential treatment for arbitrary reasons. Inclusive means that everyone, regardless of any difference, is integrated into the organization. Ethical means that actions throughout the organization conform to the Army Values and moral principles.

Fairness and Inclusiveness

8-11. A leader who uses the same set of policies and the same viewpoint in treatment of others is on the right path to building a positive climate. Although leaders should be consistent and fair in how they treat others, not everyone will be treated exactly alike. People have different capabilities and different needs, so leaders should consider some differences while ignoring irrelevant differences. Leaders need to judge certain situations according to what is important in each case. While not everyone will receive the same treatment, fair leaders will use the same set of principles and values to avoid arbitrary treatment of others.

8-12. All leaders are responsible for adhering to equal opportunity policies and preventing all forms of harassment. Creating a positive climate begins with encouraging diversity and inclusiveness.

Open and Candid Communications

8-13. Through the example they set and the leadership actions they take, good leaders will encourage open communications and candid observations. A leader who is as interested in getting others' input in advocating a position needs to encourage an environment where others feel free to contribute. An open and candid environment is a key ingredient in creating a unit that is poised to recognize and adapt to change. Approachable leaders show respect for others' opinions, even when it may represent contrary viewpoints or viewpoints out of the mainstream of thought. Some leaders specifically recognize others to provide a critical viewpoint to guard against groupthink. An open leader does not demean others and encourages input and feedback. A positive leader also remains calm and objective when receiving potentially bad news.

Learning Environment

8-14. The Army, as a learning organization, harnesses the experience of its people and organizations to improve the way it operates. Based on their experiences, learning organizations adopt new techniques and procedures that get the job done more efficiently or effectively. Likewise, they discard techniques and procedures that have outlived their purpose. Learning organizations create a climate that values and supports learning in its leaders and people. Opportunities for training and education are actively identified and supported. Leaders have direct impact on creating a climate that values learning across everyone's entire Army career. This corresponds to the same goal as lifelong learning.

Lifelong learning is the individual lifelong choice to actively and overtly pursue knowledge, the comprehension of ideas, and the expansion of depth in any area in order to progress beyond a known state of development and competency (FM 7-0).

8-15. Leaders who learn look at their experience and find better ways of doing things. It takes courage to create a learning environment. Leaders dedicated to a learning environment cannot be afraid to challenge how they and their organizations operate. When leaders question, "why do we do it this way" and find out the only reason is, "because we've always done it that way", it is time for a closer look at this process. Teams that have found a way that works may not be doing things the best way. Unless leaders are willing to question how things operate now, no one will ever know what can be done.

8-16. Leaders who make it a priority to improve their Soldiers and civilians, and the way the teams work, lead a learning organization. They use effective assessment and training methods, encourage others to reach their full potential, motivate others to develop themselves, and help others obtain training and education. An upbeat climate encourages Soldiers and civilians to recognize the need for organizational change and supports a willing attitude of learning to deal with change.

Assessing Climate

8-17. Some very definite actions and attitudes can determine climate. The members' collective sense of the group—its organizational climate—is directly attributable to the leader's values, skills, and actions. Army leaders shape the climate of the organization, no matter what the size. Conducted within 90 days of taking company command, Command Climate Surveys assist leaders in understanding the unit's climate. (See DA Pam 600-69 for information.) Answering the following questions can help assess organizational climate:

- Are clear priorities and goals set?
- Does a system of recognition, rewards, and punishments exist? Does it work?
- Do leaders know what they are doing?
- Do leaders have the courage to admit when they are wrong?
- Do leaders actively seek input from subordinates?
- Do leaders act on the feedback they have provided?
- In the absence of orders, do junior leaders have authority to make decisions when they are consistent with the commander's intent or guidance?
- Do leaders perceive high levels of internal stress and negative competition in the organization? If so, what are the options to change that situation?
- Do leaders lead by example and serve as good role models?
- Is leader behavior consistent with the Army Values?
- Do leaders lead from the front, sharing hardship when things get rough?
- Do leaders talk to the organization on a regular basis and keep people informed?

8-18. The leader's behavior has significant impact on the organizational climate. Army leaders who do the right things for the right reasons will create a healthy organizational climate. Leader behavior signals to every member of the organization what is or is not tolerated.

Dealing with Ethics and Climate

8-19. A leader is the ethical standard-bearer for the organization, responsible for building an ethical climate that demands and rewards behavior consistent with the Army Values. Other staff specialists—the chaplain, staff judge advocate, inspector general, and equal employment opportunity specialist—assist in shaping and assessing the organization's ethical climate. Regardless of all the available expert help, the ultimate responsibility to create and maintain an ethical climate rests with the leader.

8-20. Setting a good ethical example does not necessarily mean subordinates will follow it. Some may feel that circumstance justifies unethical behavior. Therefore, the leader must constantly monitor the organization's ethical climate and take prompt action to correct any discrepancies between the climate and the standard. To effectively monitor organizational climates, leaders can use a periodic Ethical Climate Assessment Survey combined with a focused leader plan of action as follows:

- Begin the plan of action by assessing the unit. Observe, interact, and gather feedback from others, or conduct formal assessments of the workplace.
- Analyze gathered information to identify what needs improvement. After identifying what needs improvement, begin developing courses of action to make the improvements.
- Develop a plan of action. First, develop and consider several possible courses of action to correct identified weaknesses. Gather important information, assess the limitations and risks associated with the various courses, identify available key personnel and resources, and verify facts and assumptions. Attempt to predict the outcome for each possible course of action. Based on predictions, select several leader actions to deal with target issues.
- Execute the plan of action by educating, training, or counseling subordinates; instituting new policies or procedures; and revising or enforcing proper systems of rewards and punishment. The organization moves towards excellence by improving substandard or weak areas and maintaining conditions that meet or exceed the standard. Finally, periodically reassess the unit to identify new matters of concern or to evaluate the effectiveness of the leader actions.

8-21. Use this process for many areas of interest and concern within the organization. It is important for subordinates to have confidence in the organization's ethical environment because much of what is necessary in war goes against the grain of societal values that individuals bring into the Army. A Soldier's conscience may say it is wrong to take human life while the mission calls for exactly that. A strong ethical climate helps Soldiers define their duty, preventing a conflict of values that may sap a Soldier's will to fight at tremendous peril to the entire team.

SGT York

Initially a conscientious objector from the Tennessee hills, Alvin C. York was drafted after America's entry into World War I and assigned to the 328th Infantry Regiment of the 82d Division, the "All Americans."

PVT York, a devout Christian, told his commander, CPT E. C. B. Danforth, that he would bear arms against the enemy—but did not believe in killing. Recognizing PVT York as a good Soldier and potential leader but unable to sway him from his convictions, CPT Danforth consulted his battalion commander, MAJ George E. Buxton, on how to handle the situation.

MAJ Buxton, a religious man with excellent knowledge of the Bible, had CPT Danforth bring PVT York to him. The major and PVT York talked at length about the Scriptures, God's teachings, about right and wrong, and just wars. Then MAJ Buxton sent PVT York home on leave to ponder and pray over the dilemma.

The battalion commander had promised to release York from the Army if he decided that he could not serve his country without sacrificing his integrity.

After two weeks of reflection and soul-searching, PVT York returned to his unit. He had reconciled his personal values with those of the Army. PVT York's decision would have great consequences for both himself and his unit.

In the morning hours of 8 October 1918 in France's Argonne Forest, now CPL York, after winning his stripes during combat in the Lorraine, would demonstrate the character and heroism that would become part of American military history.

CPL York's battalion was moving across a valley to seize a German-held rail point when a German infantry battalion, hidden on a wooded ridge overlooking the valley, opened with machine gun fire. The Americans sought cover and the attack stalled.

CPL York's platoon, reduced to 16 men, was sent to flank the enemy guns. They advanced through the woods, surprising a group of some 25 Germans. The shocked enemy troops offered only token resistance as several hidden machine guns swept the clearing with fire. The Germans immediately dropped to the ground unharmed, while nine Americans, including the platoon leader and two other corporals, fell from the hail of bullets. CPL York was the only unwounded American leader remaining.

CPL York found his platoon trapped and under fire within 25 yards of enemy machine gun pits. Instead of panicking, he began firing into the nearest enemy position, aware that the Germans would have to expose themselves to aim at him. An expert marksman, CPL York was able to hit every enemy who lifted his head over the parapet.

After CPL York shot more than a dozen, six Germans decided to charge with fixed bayonets. As the Germans ran toward him, CPL York, drawing on the instincts of a Tennessee hunter, shot the last man in the German group first, so the others would not know that they were under fire. York then shot all the assaulting Germans, moving his fire up to the front of the column. Finally, he again turned his attention to the machine gun pits. In between shots, he called at the Germans to surrender.

Although it seemed ludicrous for a lone Soldier to call on a well-entrenched enemy to surrender, the opposing German battalion commander, who had seen over 20 of his Soldiers killed, advanced and offered to surrender to CPL York if he ceased firing.

CPL York faced a daunting task. His platoon, with merely seven unwounded Soldiers, was isolated behind enemy lines with several dozen prisoners. When one American reminded York that the platoon's predicament was hopeless, he told him to be quiet.

CPL York soon moved the prisoners and his platoon toward American lines, encountering other German positions also forcing their surrender. By the time the platoon reached the edge of the valley they left just a few hours before, the hill was clear of all German machine guns. The suppressive fires on the Americans substantially reduced, the advance could continue.

CPL York returned to American lines with 132 prisoners and 35 German machine guns out of action. After delivering the prisoners, he returned to his unit. U.S. Intelligence officers later questioned the prisoners to learn that one determined American Soldier, armed with only a rifle and pistol, had defeated an entire German battalion.

For his heroic actions, CPL York was promoted to sergeant and awarded the Medal of Honor. His character, physical courage, competence, and leadership enabled him to destroy the morale and effectiveness of an entire enemy infantry battalion.

8-22. From a simply disciplinary perspective, Captain Danforth and Major Buxton could easily have ordered Private York to do his duty under threat of courts martial, or they might even have assigned him a duty away from the fighting. Instead, these two leaders appropriately addressed the Soldier's ethical concerns. Major Buxton, in particular, established the appropriate ethical climate when he showed that he, too, had wrestled with the very questions that troubled Private York. The climate the leaders created demonstrated that every person's beliefs were important and would be considered. Major Buxton established that a Soldier's duties could be consistent with the ethical framework established by his spiritual beliefs.

BUILDING TEAMWORK AND COHESION

8-23. Teamwork and cohesion are measures of climate. Willingness to engage in teamwork is the opposite of selfishness. Selfless service is a requirement for effective teamwork. To operate effectively, teams, units, and organizations need to work together for common Army Values and task and mission objectives. Leaders encourage others to work together, while promoting group pride in accomplishments. Teamwork is based on commitment to the group, which in turn is built on trust. Trust is based on expecting that others will act for the team and keep its interests ahead of their own. Leaders have to do the hard work of dealing with breaches in trust, poor team coordination, and outright conflicts. Leaders should take special care in quickly integrating new members into the team with this commitment in mind.

8-24. Leaders can shape teams to be cohesive by setting and maintaining high standards. Positive climate exists where good, consistent performance is the norm. This is very different from a climate where perfectionism is the expectation. Team members should feel that a concentrated, honest effort is appreciated even when the results are incomplete. They should feel that their leader recognizes value in every opportunity as a means to learn and to get better.

8-25. Good leaders recognize that reasonable setbacks and failures occur whether the team does everything right or not. Leaders should express the importance of being competent and motivated, but understand that weaknesses exist. Mistakes create opportunities to learn something that may not have been brought to mind.

8-26. Soldiers and Army civilians expect to be held to high but realistic standards. In the end, they feel better about themselves when they accomplish their tasks successfully. They gain confidence in leaders who help them achieve standards and lose confidence in leaders who do not know the standards or who fail to demand quality performance.

ENCOURAGING INITIATIVE

8-27. One of the greatest challenges for a leader is to encourage subordinates to exercise initiative. Soldiers and civilians who are not in leadership positions are often reluctant to recognize that a situation calls for them to accept responsibility and step forward. This could involve speaking up when the Soldier has technical knowledge or situational information that his commander does not.

8-28. Climate is largely determined by the degree to which initiative and input is encouraged from anyone with an understanding of the relevancy of the point. Leaders can set the conditions for initiative by guiding others in thinking through problems for themselves. They can build confidence in the Soldier's, or Army civilian's, competence and ability to solve problems.

DEMONSTRATING CARE FOR PEOPLE

8-29. The care that leaders show for others affects climate. Leaders who have the well-being of their subordinates in mind create greater trust. Leaders who respect those they work with will likely be shown respect in return. Respect and care can be demonstrated by simple actions such as listening patiently or ensuring that Soldiers or civilians who are deploying have their families' needs addressed. Regular sensing of morale and actively seeking honest feedback about the health of the organization also indicate care.

PREPARES SELF

8-30. To prepare for increasingly more demanding operational environments, Army leaders must invest more time on self-study and self-development than before. Besides becoming multiskilled, Army leaders have to balance the demands of diplomat and warrior. Acquiring these capabilities to succeed across the spectrum of conflicts is challenging, but critical. In no other profession is the cost of being unprepared as unforgiving, often resulting in mission failure and unnecessary casualties.

BEING PREPARED FOR EXPECTED AND UNEXPECTED CHALLENGES

8-31. Successful self-development concentrates on the key components of the leader: character, presence, and intellect. While continuously refining their ability to apply and model the Army Values, Army leaders know that in the physical arena, they must maintain high levels of fitness and health, not only to earn continuously the respect of subordinates, peers, and superiors, but also to withstand the stresses of leading and maintaining their ability to think clearly.

8-32. While physical self-development is important, leaders must also exploit every available opportunity to sharpen their intellectual capacity and knowledge in relevant domains. As addressed in Chapter 6, the conceptual components affecting the Army leader's intelligence include agility, judgment, innovation, interpersonal tact, and domain knowledge. A developed intellect helps the leader think creatively and reason analytically, critically, ethically, and with cultural sensitivity.

8-33. When faced with diverse operational settings, a leader draws on intellectual capacity, critical thinking abilities, and applicable domain knowledge. Leaders create these capabilities by frequently studying doctrine, tactics, techniques, and procedures, and by putting the information into context with personal experiences, military history, and geopolitical awareness. Here, self-development should include taking the time to learn languages, customs, belief systems, motivational factors, operational principles, and the doctrine of multinational partners and those of potential adversaries. Leaders can gain additional language skills and geopolitical awareness by seeking language schooling and assignments in specific regions of interest.

8-34. Self-development is continuous and must be pursued during both institutional and operational assignments. Successful self-development begins with the motivated individual, supplemented by a concerted team effort. Part of that team effort is quality feedback from multiple sources, including peers, subordinates, and superiors. Trust-based mentorship can also help focus self-development efforts to achieve specific professional objectives. It is important to understand that this feedback leads to establishing self-development goals and self-improvement courses of action. These courses of action are designed to improve performance by enhancing previously acquired skills, knowledge, behaviors, and experience. They further determine the potential for progressively more complex and higher-level assignments.

8-35. Generally, self-development for junior leaders is more structured and focused. The focus broadens as individuals identify their own strengths and weaknesses, determine individual needs, and become more independent. While knowledge and perspective increase with age, experience, institutional training, and operational assignments, goal-oriented self-development actions can greatly accelerate and broaden skills and knowledge. Soldiers and civilians can expect their leaders to assist in their self-development.

8-36. Civilian and military education is another important part of self-development. Army leaders never stop learning and seek out education and training opportunities beyond what is offered in required schooling or during duty assignments. To prepare for future responsibilities, Army leaders should explore off-duty education, such as available college courses that teach additional skills and broaden perspectives on life, as well as distributed learning courses on management principles or specific leadership topics.

8-37. Leaders are challenged to develop themselves and assist subordinates to acquire the individual attributes, intellectual capacities, and competencies to become the future leaders of the Army. To achieve leadership success in increasingly more complex tactical, operational, and strategic environments, leaders need to expand professional, domain knowledge and develop a keen sense of self-awareness.

EXPANDING KNOWLEDGE

8-38. Leaders prepare themselves for leadership positions through lifelong learning. Lifelong learning involves study and reflection to acquire new knowledge and to learn how to apply it when needed. Some leaders readily pick up strategies about how to learn new information faster and more thoroughly. Becoming a better learner involves several purposeful steps:

- Plan the approach to use to learn.
- Focus on specific, achievable learning goals.
- Set aside time to study.
- Organize new information as it is encountered.
- Track how learning is proceeding.

8-39. Good learners will focus on new information, what it means in relation to other information, and how it might be applied. To solidify new knowledge, try to apply it and experience what it means. Leaders need to develop and extend knowledge of tactics and operational art, technical equipment and systems, diverse cultures, and geopolitical situations. (Chapter 6 describes these domains.)

Developing Self-Awareness

8-40. Self-awareness is a component of preparing self. It is being prepared, being actively engaged in a situation and interacting with others. Self-awareness has the potential to help all leaders become better adjusted and more effective. Self-awareness is relevant for contemporary operations requiring cultural sensitivity and for a leader's adaptability to inevitable environmental change.

8-41. Self-awareness enables leaders to recognize their strengths and weaknesses across a range of environments and progressively leverage strengths to correct these weaknesses. To be self-aware, leaders must be able to formulate accurate self-perceptions, gather feedback on others' perceptions, and change their self-concept as appropriate. Being truly self-aware ultimately requires leaders to develop a clear, honest picture of their capabilities and limitations.

Self-awareness is being aware of oneself, including one's traits, feelings, and behaviors.

8-42. As a given situation changes, so must a leader's assessment of abilities and limitations in order to adapt. Every leader has the ability to be self-aware. Competent leaders understand the importance of self-awareness and work to develop it.

8-43. In contrast, leaders who lack self-awareness are often seen as arrogant and disconnected from their subordinates. They may be technically competent but lack of awareness as to how they are seen by subordinates. This may also obstruct learning and adaptability, which in turn, keeps them from creating a positive work climate and a more effective organization. Self-aware leaders understand the variety of Soldiers and civilians on their team. They sense how others react to their actions, decisions, and image.

8-44. Self-aware leaders are open to feedback and actively seek it. A leader's goal in obtaining feedback is to develop an accurate self-perception by understanding other people's perceptions. Many leaders have successfully used a multisource assessment and feedback method to gain insight. A multisource assessment is a formal measure of peer, subordinate, superior, and self-impressions of a single individual. It may provide critical feedback and insights that are otherwise not apparent.

8-45. The Army's after-action review (AAR) process is a well-used awareness tool. Its purpose is to help units and individuals identify their strengths and weaknesses. A productive self-review occurs when one examines his or her self and becomes conscious of one's own behavior and interactions with others.

8-46. Leaders should also seek out others to help them make sense of their experiences. Talking with coaches, friends, or other trusted individuals can provide valuable information. Most, but not all Army leaders, find a mentor whom they trust to provide honest feedback and encouragement.

8-47. It is important to realize that feedback does not have to be gathered in formal counseling, survey, or sensing sessions. Some of the best feedback comes from simply sitting down and informally talking with Soldiers and civilians. Many commanders have gained valuable information about themselves from merely eating a meal in the dining facility with a group of Soldiers and asking about unit climate and training.

8-48. Self-aware leaders analyze themselves and ask hard questions about experiences, events, and their actions. They should examine their own behavior seriously. Competent and confident leaders make sense of their experience and use it to learn more about themselves. Journals and AARs are valuable tools to help gain an understanding of one's past experiences and reactions to the changes in the environment. Self-critique can be as simple as posing questions about one's own behavior, knowledge, or feelings. It can be as formal as answering a structured set of questions about a high profile event. Critical questions include—

- What happened?
- How did I react?

- How did others react and why?
- What did I learn about myself based on what I did and how I felt?
- How will I apply what I learned?

8-49. In the rapidly changing environment of both the current and future force, leaders are faced with unfamiliar and uncertain situations. For any leader, self-awareness is a critical factor in making accurate assessments of the changes in the environment and their personal capabilities and limitations to operate in that environment. Self-awareness helps leaders translate prior training to a new environment and seek out new information when the situation requires. Self-aware leaders are better informed and able to determine what needs to be learned and what assistance they need to seek out to handle a given situation.

8-50. Adjusting one's thoughts, feelings, and actions based on self-awareness is called self-regulation. It is the proactive and logical follow-up to self-awareness. When leaders determine a gap from actual "self" to desired "self," they should take steps to close the gap. Leaders can seek new perspectives about themselves and turn those perspectives into a leadership advantage. Because leaders cannot afford to stop learning, they seek to improve and grow. Becoming more self-aware is not something that happens automatically. Competent and confident leaders seek input and improvements over the entire span of their careers.

DEVELOPS OTHERS

...[G]ood NCOs are not just born—they are groomed and grown through a lot of hard work and strong leadership by senior NCOs.

William A. Connelly
Sergeant Major of the Army (1979-1983)

8-51. Leader development is a deliberate, continuous, sequential, and progressive process grounded in the Army Values. It grows Soldiers and civilians into competent and confident leaders capable of directing teams and organizations to execute decisive action. Leader development is achieved through the lifelong synthesis of the knowledge, skills, and experiences gained through institutional training and education, organizational training, operational experience, and self-development.

8-52. Leader development takes into consideration that military leaders are inherently Soldiers first and must be technically and tactically proficient as well as adaptive to change. Army training and leader development therefore centers on creating trained and ready units, led by competent and confident leaders. The concept acknowledges an important interaction that trains Soldiers now and develops leaders for the future.

8-53. The three core domains that shape the critical learning experiences throughout Soldiers' and leaders' careers are—

- Institutional training.
- Training, education, and job experience gained during operational assignments.
- Self-development.

8-54. These three domains interact by using feedback and assessment from various sources and methods. Although leader development aims at producing competent leadership at all levels, it recognizes small unit leaders must reach an early proficiency to operate in widely dispersed areas in combined arms teams. The Army increasingly requires proficient small unit leaders capable of operating in widely dispersed areas and/or integrated with joint, multinational, special operations forces as well as nongovernmental agencies. These leaders must be self-aware and adaptive, comfortable with ambiguity, able to anticipate possible second- and third-order effects, and be multifunctional to exploit combined arms integration.

8-55. To that end, the Army leverages leader development education (professional military education and the Civilian Education System), ensuring the best mix of experiences and operational assignments supported by resident and distributed education. The effort requires improved individual assessment and feedback and increased development efforts at the organizational level in the form of mentoring, coaching, and counseling, as well as picking the right talent for specific job assignments. The purpose of the increased developments efforts is to instill in all Soldiers and leaders the desire and drive to update their professional knowledge and competencies, thus improving current and future Army leaders' abilities to master the challenges of full spectrum operations.

8-56. Leader development also requires organizational support. A commander or other designated leader has the responsibility to develop others for better performance in their current and future positions. There are specific actions that leaders can take to personalize leader development in their organization.

ASSESSING DEVELOPMENTAL NEEDS

8-57. The first step in developing others is to understand how they may be developed best; what areas are already strong and what areas should be stronger. Leaders who know their subordinates will have an idea where to encourage them to develop. New subordinates can be observed under different task conditions to identify strengths and weaknesses to see how quickly they pick up new information and skills.

8-58. Leaders often conduct an initial assessment before they take over a new position. They ask themselves questions: how competent are new subordinates? what is expected in the new job? Leaders review the organization's standing operating procedure and any regulations that apply as well as status reports and recent inspection results. They meet with the outgoing leader and ask for an assessment and meet with key people outside the organization. Leaders listen carefully as everyone sees things through personal filters. .They reflect and realize initial that their impressions may still be off base. Good leaders update in-depth assessments with assumption of new duty positions since a thorough assessment assists in implementing changes gradually and systematically without causing damaging organizational turmoil.

8-59. To objectively assess subordinates, leaders do the following:

- Observe and record subordinates' performance in the core leader competencies.
- Determine if the performances meet, exceed, or fall below expected standards.
- Tell subordinates what was observed and give an opportunity to comment.
- Help subordinates develop an individual development plan (IDP) to improve performance.

8-60. Good leaders provide honest feedback to others, discussing strengths and areas for improvement. Effective assessment results in an IDP designed to correct weaknesses and sustain strengths. Here is what is required to move from planning to results:

- Design the individual development plan together, but let the subordinate take the lead.
- Agree on the required actions to improve leader performance in the core leader competencies. Subordinates must buy into this plan if it is going to work.
- Review the plan frequently, check progress, and modify the plan if necessary.

DEVELOPING ON THE JOB

8-61. The best development opportunities often occur on the job. Leaders who have an eye for developing others will encourage growth in current roles and positions. How a leader assigns tasks and duties can serve as a way to direct individual Soldiers or civilians to extend their capabilities. The Army civilian intern program is an excellent example of this type of training. Feedback from a leader during routine duty assignments can also direct subordinates to areas where they can focus their development. Some leaders constantly seek new ways to re-define duties or enrich a job to prepare subordinates for additional responsibilities in their current position or next assignment. Cross training on tasks provides dual benefits of building a more robust team and expanding the skill set of team members. Challenging subordinates with different job duties is a good way to keep them interested in routine work.

SUPPORTING PROFESSIONAL AND PERSONAL GROWTH

8-62. Preparing self and subordinates to lead aims at the goal of developing multiskilled leaders—leader pentathletes. The adaptable leader will more readily comprehend the challenges of a constantly evolving strategic environment, demanding not only warfighting skills, but also creativity and a degree of diplomacy combined with multicultural sensitivity. To achieve this balance, the Army creates positive learning environments at all levels to support its lifelong learning strategy.

8-63. As a lifelong learning institution, the Army addresses the differences between operations today and in the future and continuously develops enhanced training and leader development capabilities. Army leaders who look at their experiences and learn from them will find better ways of doing things. It takes openness and imagination to create an effective organizational learning environment. Do not be afraid to make mistakes. Instead, stay positive and learn from those mistakes. Leaders must remain confident in their own and their subordinates' ability to make learning the profession of arms a lifelong commitment. This attitude will allow growth into new responsibilities and adapt to inevitable changes. French military theorist Ardant Du Picq stressed the importance of learning:

The instruments of battle are valuable only if one knows how to use them….

8-64. Leaders who have the interest of others and the organization in mind will fully support available developmental opportunities, nominate and encourage subordinates for those opportunities, help remove barriers to capitalize on opportunities, and see that the new knowledge and skills can be reinforced once they are back on the job.

HELPING PEOPLE LEARN

8-65. In any developmental relationship, the leader can adopt special ways to help others learn. It is the leader's responsibility to help subordinates to learn. Certain instructions clearly help people learn. Explain why a subject is important. Leaders show how it will help individuals and the organization perform better and actively involve subordinates in the learning process. For instance, never try to teach someone how to drive a vehicle with classroom instruction alone. Ultimately, the person has to get behind the wheel. To keep things interesting, keep lectures to a minimum and maximize hands-on training.

8-66. Learning from actual experience is not always possible. Leaders cannot have every experience in training. They substitute for that by taking advantage of what others have learned and getting the benefit without having the personal experience. Leaders should also share their experiences with subordinates during counseling, coaching, and mentoring, such as combat veterans sharing experiences with Soldiers who have not been to war.

COUNSELING, COACHING AND MENTORING

Soldiers learn to be good leaders from good leaders.

Richard A. Kidd
Sergeant Major of the Army (1991-1995)

8-67. Leaders have three principal ways of developing others. They can provide others with knowledge and feedback through counseling, coaching, and mentoring:

- Counseling—occurs when a leader, who serves as a subordinate's designated rater, reviews with the subordinate his demonstrated performance and potential, often in relation to a programmed performance evaluation.
- Coaching—the guidance of another's person's development in new or existing skills during the practice of those skills.
- Mentoring—a leader with greater experience than the one receiving the mentoring provides guidance and advice; it is a future-oriented developmental activity focused on growing in the profession.

Counseling

8-68. Counseling is central to leader development. Leaders who serve as designated raters have to prepare their subordinates to be better Soldiers or civilians. Good counseling focuses on the subordinate's performance and problems with an eye toward tomorrow's plans and solutions. The subordinate is expected to be an active participant who seeks constructive feedback. Counseling cannot be an occasional event but should be part of a comprehensive program to develop subordinates. With effective counseling, no evaluation report—positive or negative—should be a surprise. A consistent counseling program includes all subordinates, not just the people thought to have the most potential.

Counseling is the process used by leaders to review with a subordinate the subordinate's demonstrated performance and potential.

8-69. During counseling, subordinates are not passive listeners but active participants in the process. Counseling uses a standard format to help mentally organize and isolate relevant issues before, during, and after the counseling session. During counseling, leaders assist subordinates to identify strengths and weaknesses and create plans of action. To make the plans work, leaders actively support their subordinates throughout the implementation and assessment processes. (See Appendix B for a detailed discussion on counseling.) Subordinates invest themselves in the process by being forthright in their willingness to improve and being candid in their assessment and goal setting.

8-70. The three types of counseling are—

- Event counseling.
- Performance counseling.
- Professional growth counseling.

Event Counseling

8-71. Event counseling covers a specific event or situation. It may precede events such as going to a promotion board or attending a school. It may also follow events such as an exceptional duty performance, a performance problem, or a personal problem. Event counseling is also recommended for reception into a unit or organization, for crises, and for transition from a unit or separation from the Army.

Performance Counseling

8-72. Performance counseling is the review of a subordinate's duty performance during a specified period. The leader and the subordinate jointly establish performance objectives and clear standards for the next counseling period. The counseling focuses on the subordinate's strengths, areas to improve, and potential. Effective counseling includes providing specific examples of strengths and areas needing improvement and providing guidance on how subordinates can improve their performance. Performance counseling is required under the officer, noncommissioned officer (NCO), and Army civilian evaluation reporting systems.

Professional Growth Counseling

8-73. Professional growth counseling includes planning for the accomplishment of individual and professional goals. It has a developmental orientation and assists subordinates in identifying and achieving organizational and individual goals. Professional growth counseling includes a review to identify and discuss the subordinate's strengths and weaknesses and the creation of an IDP. The plan builds on existing strengths to overcome weaknesses.

8-74. A part of professional growth counseling is a discussion characterized as a "pathway to success." It establishes short- and long-term goals for the subordinate. These goals may include opportunities for civilian or military schooling, future duty assignments, special programs, or reenlistment options. Leaders help develop specific courses of action tailored to each individual. For example, during required career field counseling for lieutenants and captains, raters and senior raters, together with the rated officer, determine how the rated officer's skills and talents best fit the needs of the Army. They allow special consideration to the rated officer's preferences and abilities.

Approaches to Counseling

8-75. Inexperienced leaders are sometimes uncomfortable when confronting a subordinate who is not performing to standard. Counseling is not about leader comfort; it is about correcting the performance or developing the character of a subordinate. To be effective counselors, Army leaders must demonstrate certain qualities: respect for subordinates, self-awareness, cultural awareness, empathy, and credibility.

8-76. One challenging aspect of counseling is selecting the proper approach for a specific situation. To counsel effectively, the technique used must fit the situation. Some cases may only require giving information or listening. A subordinate's improvement may call for just a brief word of praise. Other situations may require structured counseling followed by specific plans for actions. An effective leader approaches each subordinate as an individual. Counseling includes nondirective, directive, and combined approaches. The major difference between the approaches is the degree to which the subordinate participates and interacts during a counseling session.

8-77. The **nondirective approach** is preferred for most counseling sessions. Leaders use their experiences, insight and judgment to assist subordinates in developing solutions. Leaders partially structure this type of counseling by telling the subordinate about the counseling process and explaining expectations.

8-78. The **directive approach** works best to correct simple problems, make on-the-spot corrections, and correct aspects of duty performance. When using the directive style, the leader does most of the talking and tells the subordinate what to do and when to do it. In contrast to the nondirective approach, the leader directs a course of action for the subordinate.

8-79. In the **combined approach**, the leader uses techniques from both the directive and nondirective approaches, adjusting them to articulate what is best for the subordinate. The combined approach emphasizes the subordinate's planning and decision-making responsibilities.

Coaching

8-80. While a mentor or counselor generally has more experience than the person being supported does, coaching relies primarily on teaching and guiding to bring out and enhance the capabilities already present. From its original meaning, coaching refers to the function of helping someone through a set of tasks. Those being coached may, or may not, have appreciated their potential. The coach helps them understand their current level of performance and instructs them how to reach the next level of knowledge and skill.

8-81. When compared to counseling and mentoring, coaching is a development technique that tends to be used for a skill and task-specific orientation. Coaches should possess considerable knowledge in the specific area in which they coach others.

8-82. An important aspect of coaching is identifying and planning for short- and long-term goals. The coach and the person being coached discuss strengths, weaknesses, and courses of action to sustain or improve. Coaches use the following guidelines:

- **Focus Goals**: This requires the coach to identify the purpose of the coaching session. Expectations of both the person being coached and the coach need to be discussed. The coach communicates to the individual the developmental tasks for the coaching session, which can incorporate the results of the individual's multisource assessment and feedback survey.
- **Clarify the Leader's Self-Awareness**: The coach works directly with the leader to define both strengths and developmental needs. During this session, the coach and the leader communicate perceived strengths, developmental needs, and focus areas to improve leader performance. Both the coach and the individual agree on areas of developmental needs.
- **Uncover Potential**: The coach facilitates self-awareness of the leader's potential and the leader's developmental needs by guiding the discussion with questions. The coach actively listens to how the leader perceives his potential. The aim is to encourage the free flow of ideas. The coach also assesses the leader's readiness to change and incorporates this into the coaching session.
- **Eliminate Developmental Barriers**: The coach identifies developmental needs with the leader and communicates those areas that may hinder self-development. It is during this step that the coach helps the individual determine how to overcome barriers to development and how to implement an effective individual development plan to improve the leader's overall performance. The coach helps the leader identify potential sources of support for implementing an action plan.
- **Develop Action Plans and Commitment**: The coach and the individual develop an action plan defining specific actions that can improve the leader's performance within a given period. The coach utilizes a developmental action guide to communicate those self-directed activities the leader can accomplish on his own to improve his performance within a particular competency.

- **Follow-Up**: After the initial coaching session, there should be a follow up as part of a larger transition. After the initial coaching, participants should be solicited for their feedback concerning the effectiveness of the assessment, the usefulness of the information they received, and their progress towards implementing their IDP. The responsibility for follow-up coaching, further IDP development, and IDP execution is usually the responsibility of the unit chain of command. Leaders in the chain of command who provide coaching have a profound impact on the development of their subordinate leaders. They are the role models and present subordinates with additional information and incentives for self-development. Leaders who coach provide frequent informal feedback and timely, proactive, formal counseling to regularly inspire and improve their subordinates.

Mentoring

8-83. Future battlefield environments will place additional pressures on developing leaders at a rapid pace. To help these leaders acquire the requisite abilities, the Army relies on a leader development system that compresses and accelerates development of professional expertise, maturity, and conceptual and team building skills. Mentoring is a developmental tool that can effectively support many of these learning objectives. It is a combat multiplier because it boosts positive leadership behaviors on a voluntary basis.

8-84. It is usually unnecessary for leaders to have the same occupational or educational background as those they are coaching or counseling. In comparison, mentors generally specialize in the same specific area as those being mentored. Mentors have likely experienced what their protégés and mentees are experiencing, or are going to experience. Consequently, mentoring relationships tend to be occupation and/or domain specific, with the mentor having expertise in the particular areas they are assisting in, but without the requirement to have the same background. Mentoring focuses primarily on developing a less experienced leader for the future.

Mentorship is the voluntary developmental relationship that exists between a person of greater experience and a person of lesser experience that is characterized by mutual trust and respect (AR 600-100).

8-85. The focus of mentorship is the voluntary mentoring that goes beyond the chain of command. Mentorship is generally characterized by the following:

- Mentoring takes place when the mentor provides a less experienced leader with advice and counsel over time to help with professional and personal growth.
- The developing leader often initiates the relationship and seeks counsel from the mentor. The mentor takes the initiative to check on the well-being and development of that person.
- Mentorship affects both personal development (maturity, interpersonal, and communication skills) as well as professional development (technical and tactical knowledge and career path knowledge).
- Mentorship helps the Army maintain a highly competent set of leaders.
- The strength of the mentoring relationship is based on mutual trust and respect. The mentored carefully consider assessment, feedback, and guidance; these considerations become valuable for the growth that occurs.

8-86. Contrary to common belief, mentoring relationships are not confined to the superior-subordinate relationship. They may also be found between peers and notably between senior NCOs and junior officers. This relationship can occur across many levels of rank. In many circumstances, this relationship extends past the point where one or the other has left the chain of command.

8-87. Supportive mentoring occurs when a mentor does not outrank the person being mentored, but has more extensive knowledge and experience. Early in their careers, young officers are paired with senior experienced NCOs. The relationship that frequently comes from this experience tends to be instrumental in the young officer's development. Often, officers will recognize that the noncommissioned officer in their first or second assignment was a critical mentor with a major impact on their development.

8-88. Individuals must be active participants in their developmental process. They must not wait for a mentor to choose them but have responsibility to be proactive in their own development. Every Army officer, NCO, Soldier, and civilian should identify specific personal strengths, weaknesses, and areas in need of improvement. Each individual should then determine a developmental plan to correct these deficiencies. Some strategies that may be used are to—

- Ask questions and pay attention to experts.
- Read and study.
- Watch those in leadership positions.
- Find educational opportunities (civilian, military, and correspondence).
- Seek and engage in new and varied opportunities.

8-89. Soldiers can increase their chances of being mentored by actively seeking performance feedback and by adopting an attitude of lifelong learning. These self-development actions help set the stage for mentoring opportunities. Soldiers who seek feedback to focus their development, coupled with dedicated, well-informed mentors, will be the foundation for embedding the concepts of lifelong learning, self-development, and adaptability into the Army's culture.

8-90. While mentoring is generally associated with improving duty-related performance and growth, it does not exclude a spiritual dimension. A chaplain or other spiritually trained or enlightened individual may play a significant role in helping individuals cope with stress and find better professional balance and purpose.

BUILDING TEAM SKILLS AND PROCESSES

The cohesion that matters on the battlefield is that which is developed at the company, platoon, and squad levels....

General Edward C. Meyer
Chief of Staff, Army (1979-1983)

8-91. The national cause, the purpose of the mission, and many other concerns may not be visible from the Soldier's perspective on the battlefield. Regardless of larger issues, Soldiers perform for the other people in the squad or section, for others on the team or crew, for the person on their right or left. It is a fundamental truth, born from the Warrior Ethos. Soldiers get the job done because they do not want to let their friends down. Similarly, Army civilians feel part of the installation and organizational team and want to be winners.

8-92. Developing close teams takes hard work, patience, and interpersonal skill on the part of the leader. It is a worthwhile investment because good teams complete missions on time with given resources and a minimum of wasted effort. In combat, cohesive teams are the most effective and take the fewest casualties.

Characteristics of Teams

8-93. The hallmarks of close teams include—

- Trusting each other and being able to predict what each other will do.
- Working together to accomplish the mission.
- Executing tasks thoroughly and quickly.
- Meeting and exceeding the standard.
- Thriving on demanding challenges.
- Learning from their experiences and developing pride in their accomplishments.

8-94. The Army as a team includes many members who are not Soldiers. The contributions made by countless Army civilians, contractors, and multinational personnel in critical support missions during Operation Desert Storm, the Balkans, and the War on Terrorism are often forgotten. In today's logistic-heavy operational environments, many military objectives could not be achieved without the dedicated support of the Army's hard-working civilian team members.

8-95. Within a larger team, smaller teams may be at different stages of development. For instance, members of First Squad may be accustomed to working together. They trust one another and accomplish the mission, usually exceeding the standard without wasted effort. Second Squad in the same platoon just received three new Soldiers and a team leader from another company. As a team, Second Squad is less mature and it will take them some time to get up to the level of First Squad. Second Squad's new team members have to learn how things work. First, they have to feel like members of the team. Subsequently, they must learn the standards and the climate of their new unit and demonstrate competence before other members really accept them. Finally, they must practice working together. Leaders can best oversee the integration process if they know what to expect.

8-96. Competent leaders are sensitive to the characteristics of the team and its individual members. Teams develop differently and the boundaries between stages are not hard and fast. The results can help determine what to expect of the team and what is needed to improve its capabilities.

Stages of Team Building

8-97. Figure 8-1 lists actions that pull a team together. Teams do not come together by accident. Leaders must guide them through three developmental stages:

- Formation.
- Enrichment.
- Sustainment.

Formation Stage

8-98. Teams work best when new members quickly feel a part of the team. The two critical steps of the formation stage—reception and orientation—are dramatically different in peace and war. In combat, a good sponsorship process can literally make the difference between life and death for new arrivals and to the entire team.

8-99. Reception is the leader's welcome to the organization. Time permitting; it should include a handshake and personal introduction. The orientation stage begins with meeting other team members, learning the layout of the workplace, learning the schedule, and generally getting to know the environment. In combat, leaders may not have much time to spend with new members. In this case, a sponsor is assigned to new arrivals. That person will help them get oriented until they "know the ropes."

8-100. In combat, Army leaders have countless things to worry about and the mental state of new arrivals might seem low on the list. If Soldiers cannot fight, the unit will suffer needless casualties and may ultimately fail to complete the mission.

8-101. Discipline and shared hardships pull people together in powerful ways. SGT Alvin C. York described cohesion in this clear and simple way:

> *The war brings out the worst in you. It turns you into a mad, fightin' animal, but it also brings out something else, something I jes don't know how to describe, a sort of tenderness and love for the fellows fightin' with you.*

FORMATION STAGE	Subordinate Actions	Leader & Organizational Actions
General Team Building	• Learn about team purpose, tasks, and standards. • Learn about leaders and other members. • Achieve belonging and acceptance.	• Design effective reception and orientation. • Create learning experiences. • Communicate expectations. • Listen to and care for subordinates. • Reward positive contributions. • Set example.
Team Building for Deployments	• Adjust to uncertainty across the spectrum of conflict. • Cope with fear of unknown injury and death. • Adjust to separation from home and family.	• Talk with each Soldier. • Reassure with calm presence. • Communicate vital safety tips. • Provide stable situation. • Establish buddy system. • Help Soldiers deal with immediate problems.
ENRICHMENT STAGE		
General Team Building	• Trust leaders and other members. • Cooperate with team members. • Share information. • Accept the way things are done. • Adjust to feelings about how things ought to be done.	• Trust and encourage trust. • Reinforce desired group norms. • Establish clear lines of authority. • Establish individual and unit goals. • Identify and grow leaders. • Train as a unit for mission. • Build pride through accomplishment.
Team Building for Deployments	• Demonstrate competence. • Become a team member. • Learn about the threat. • Learn about the area of operations. • Avoid life-threatening mistakes.	• Demonstrate competence. • Prepare as a unit for operations. • Know the Soldiers. • Provide stable unit climate. • Emphasize safety for improved readiness.
SUSTAINMENT STAGE		
General Team Building	• Trust others. • Share ideas and feelings freely. • Assist other team members. • Sustain trust and confidence. • Share missions and values.	• Demonstrate trust. • Focus on teamwork, training, and maintaining. • Respond to subordinate problems. • Devise more challenging training. • Build pride and spirit.
Team Building for Deployments	• Adjust to continuous operations. • Cope with casualties. • Adjust to enemy actions. • Overcome boredom. • Avoid rumors. • Control fear, anger, despair, and panic.	• Observe and enforce sleep discipline. • Sustain safety awareness. • Inform Soldiers. • Know and deal with Soldiers' perceptions. • Keep Soldiers productively busy. • Use in-process reviews (IPRs) and after-action reviews (AARs). • Act decisively in face of panic.

Figure 8-1. Stages of team building

Enrichment Stage

8-102. New teams and new team members gradually move from questioning everything to trusting themselves, their peers, and their leaders. Leaders learn to trust by listening, following up on what they hear, establishing clear lines of authority, and setting standards. By far the most important thing a leader does to strengthen the team is training. Training takes a group of individuals and molds them into a team while preparing them to accomplish their missions. Training occurs during all three stages of team building, but is particularly important during enrichment. It is at this point that the team is building collective proficiency.

Sustainment Stage

8-103. During this stage, members identify with "their team." They own it, have pride in it, and want the team to succeed. At this stage, team members will do what is necessary without being told. Every new mission gives the leader a chance to strengthen the bonds and challenge the team to reach for new heights of accomplishment. The leader develops his subordinates because he knows they will be tomorrow's team leaders. The team should continuously train so that it maintains proficiency in the collective and individual tasks it must perform to accomplish its missions.

Chapter 9

Achieving

9-1. Leadership builds effective organizations. Effectiveness is most directly related to the core leader competency of **getting results**. From the definition of leadership, **achieving** is focused on accomplishing the mission. Mission accomplishment is a goal that must co-exist with an extended perspective towards maintaining and building up the organization's capability for the future. Achieving begins in the short term by setting objectives. In the long term, achieving based on clear vision requires getting results in pursuit of those objectives. **Getting results** is focused on structuring what needs to be done so results are consistently produced. This competency focuses on the organization of how to achieve those results.

9-2. Getting results embraces all actions to get the job done on time and to standard:

- Providing direction, guidance, and clear priorities involves guiding teams in what needs to be done and how.
- Developing and executing plans for mission and task accomplishment involves anticipating how to carry out what needs to be done, managing the resources used to get it done, and conducting the necessary actions.
- Accomplishing missions consistently and ethically involves using monitoring to identify strengths and correct weaknesses in organizational, group, and individual performance.

PROVIDING DIRECTION, GUIDANCE, AND PRIORITIES

It is in the minds of the commanders that the issue of battle is really decided.

Sir Basil H. Liddell Hart
Thoughts on War (1944)

9-3. As leaders operate in larger organizations, their purpose, direction, guidance, and priorities typically become forward-looking and wider in application. Direct level leaders and small unit commanders usually operate with less time for formal planning than organizational and strategic level leaders. Although leaders use different techniques for guidance depending on the amounts of time and staff available, the basics are the same. The leader provides guidance so subordinates and others understand the goals and priorities.

9-4. Whether operating with an infantry squad, a finance section, or an engineer team, leaders will match their teams, units, or organizations to the work required. Most work is defined by standard operating procedures and tasks assigned to groups. As new tasks develop and priorities change, assignments will differ. In higher-level positions, commanders and directors have others to help perform these assignment and prioritization functions. Higher-level organizations also have procedures such as running estimates and the military decisionmaking process to define and synchronize planning activities (see FM 5-0).

9-5. Leaders should provide guidance from both near-term and long-term perspectives. Good leaders make thoughtful trade-offs between providing too much or too little guidance. A near-term focus is based on critical actions that must be accomplished immediately. In contrast, by delegating as much as possible, leaders prepare others to handle future tasks competently and are available for higher-level coordination.

9-6. When tasks are difficult, adaptive leaders identify and account for the capabilities of the team. Some tasks will be routine and will require little clarification from the leader, while others will present new challenges for the knowledge and experience that the team has. When a new task is undertaken for the first time working with a new group, leaders are alert to group organization, their capabilities, and their commitment to the task.

9-7. Leaders should provide frequent feedback as an embedded, natural part of the work. While it is important to have set periods for developmental performance counseling, it is also important to provide feedback on a regular basis. Making feedback part of the normal performance of work is a technique leaders use to guide how duties are accomplished.

9-8. Often the most challenging of the leader's jobs is to identify and clarify conflicts in followers' roles and responsibilities. Good communication techniques with brief backs are useful for identifying conflicts. Role differences may arise during execution and should be resolved by the leader as they occur.

9-9. Good guidance depends on understanding how tasks are progressing, so the leader knows if and when to provide clarification. Most workers have a desire to demonstrate competence in their work, so leaders need to be careful that they do not reduce this drive.

DEVELOPING AND EXECUTING PLANS

A plan is a proposal for executing a command decision or project. Planning is the means by which the leader or commander envisions a desired outcome and lays out effective ways of achieving it. In the plan, the leader communicates his vision, intent and decisions and focuses his subordinates on the results he expects to achieve.

FM 3-0

9-10. In daily peacetime or combat training and operations, a leader's primary responsibility is to help the organization function effectively. The unit must accomplish the mission despite any surrounding chaos. This all begins with a well thought out plan and thorough preparation.

PLANNING

9-11. Leaders use planning to ensure that an approach for reaching goals will be practical. Planning reduces confusion, builds subordinates' confidence in themselves and their organization, and allows flexibility to adjust to changing situations. Good planning boosts shared understanding and ensures that a mission is accomplished with a minimum of wasted effort and fewer casualties in combat. FM 6-0 discusses the different types of plans in more detail.

Considering Intended and Unintended Consequences

9-12. Plans and the actions taken in those plans will most likely have unintended, as well as intended, consequences. Leaders should think through what they can expect to happen because of a plan or course of action. Some decisions may set off a chain of events that are contrary to the desired effects. Intended consequences are the anticipated results of a leader's decisions and actions. Unintended consequences arise from unplanned events that affect the organization or accomplishment of the mission. Intended and unintended consequences can best be addressed during wargaming and rehearsals that are critical during planning. The aim of wargaming and rehearsals is to reduce the unintended consequences to as few as possible.

9-13. Even lower-level leaders' actions may have effects well beyond what they expect. Consider the case of a sergeant whose team is operating a roadblock as part of peace enforcement. Early one morning, a truckload of civilians appears racing toward the roadblock. In the half-light, the noncommissioned officer (NCO) in charge of the checkpoint cannot tell if the objects in the passengers' hands are weapons or harmless farm tools—while the driver seems intent on proceeding without stopping. In the space of a few seconds, the NCO must decide whether to order the team to fire on the truck.

9-14. If the sergeant orders the team to fire to force the truck to stop, that decision can easily have international and strategic consequences. If any innocent civilians are killed, chances are good the chain of command and the outside world will know about the incident in a few short hours. The decision is tough for another reason: If the sergeant does not order the team to fire and the civilians turn out to be armed insurgents, the team may suffer unnecessary casualties.

9-15. Ultimately, the sergeant must act as the leader in charge. Leaders who think through the consequences of possible actions and understand the commander's intent, mission priorities, and rules of engagement are usually prepared to take the right steps. The intended consequences are obvious in the roadblock example: to control access of people and to eliminate entry of explosives, weapons, and contraband.

9-16. Thinking ahead about intended consequences and beyond to unintended consequences serves to sharpen what is important in the planning process. In this checkpoint example, the intended consequences of conducting an effective and secure operation might be foiled if vehicle drivers are not properly warned of the checkpoint. Are there signs in the appropriate language to demand a slow approach speed? Are there speed bumps to force a slow-go approach? Is the traffic properly funneled to prevent a bypass or escape? If these and other measures are not considered and implemented, the unintended consequences could include accidentally engaging vehicles carrying innocent civilians because of possible driver reactions that might be misinterpreted as hostile behavior.

9-17. Sometimes consequences are not direct and immediate. These types of consequences are referred to as second- and third-order effects. These effects can be intended or unintended. In the checkpoint example, the second-order effect of setting up a checkpoint may be to reduce the amount of civilian traffic in the area. The third-order effect may be to slow down the restoration of commerce in the area or the checkpoint may provide insurgents a target where local civilians gather at predictable periods during the day. An unintended consequence of the NCO's decision to fire on a speeding truckload of civilians may cause a second-order effect of local outrage. A possible third-order effect is it may cause an international incident. However, second- and third-order effects should not be the basis for hindering initiative or doing the right thing.

Reverse Planning

9-18. Reverse planning is a specific technique used to ensure that a concept leads to the intended end state. It begins with the goal or desired mission outcome in mind. The start point is the question: "Where do I want to end up?" From there, think and work the plan backwards to the current situation. While following the thought process from projected goal to current position, establish the basics steps along the way and determine the who, what, when, where, and why to accomplish the goal.

9-19. While planning, leaders consider the amount of time needed to coordinate and conduct each step. For instance, a tank platoon sergeant whose platoon has to spend part of a field exercise on the firing range might have to arrange for refueling at the range. No one explicitly said to refuel at the range, but the platoon sergeant knows it needs to happen, given the heavy fuel consumption of M1A2 tanks. Consequently, the platoon sergeant must think through the steps from the last to the first: (1) when the refueling must be complete, (2) how long the refueling will take, (3) how long it takes the refueling unit to get set up, and (4) when the refueling vehicles must report to the range.

9-20. After determining what must happen on the way to the goal, leaders put the tasks in logical sequence, set clear priorities, and determine a realistic time line. They examine all steps required in the order they will occur and if time permits solicit input from subordinates. Experienced subordinates can often provide a valuable reality check for the plan. Subordinates' input also shows their part-ownership of the plan. Positively contributing builds trust while boosting their self-confidence and will to succeed.

PREPARING

9-21. Preparation complements planning. Doctrinally, preparation for combat includes plan refinement, rehearsals, reconnaissance, coordination, inspections, and movement. See FM 3-0 and FM 6-0 for more information. In all cases, preparation includes detailed coordination with other organizations involved or affected by the operation or project. In the case of a nontactical requirement, preparation may include ensuring the necessary facilities (for example, hospitals, labs, maintenance shops) and other resources (for example, firefighters, police, and other first responders) are available to support the mission.

9-22. A rehearsal is a critical element of preparation. It allows everyone involved in a mission to develop a mental picture of responsibilities and what should happen. It helps the team synchronize operations at times and places critical to successful mission accomplishment. FM 6-0 features a detailed appendix on rehearsals. Rehearsing key combat actions allows subordinates to see how things are supposed to work and builds confidence in the plan. Even a simple walk-through helps leaders visualize who is supposed to be at a specific location to perform a coordinated action at a certain time. Leaders can see how things might unfold, what might go wrong, and how the plan could change to adjust for intended or unintended consequences.

EXECUTING

...[A] good plan violently executed now is better than a perfect plan next week.

General George S. Patton, Jr.
War As I Knew It (1947)

9-23. Successful execution of a plan is based on all the work that has gone before. Executing for success requires situational understanding, supervising task completion, assessing progress, and implementing required execution or adjustment decisions (FM 6-0).

9-24. Executing in combat means putting a plan into action by applying combat power to accomplish the mission and using situational understanding to assess progress and make execution and adjustment decisions. In combat, leaders strive to effectively integrate and synchronize all elements of the joint and combined arms team as well as nonmilitary assets. The goal is to assign specific tasks or objectives to the most capable organization and empowering its leaders to execute and exercise initiative within the given intent.

9-25. Planning execution involves awareness of whether critical tasks are being accomplished on the way to mission completion. Good leaders know which of the most important parts of the mission to check. Knowing from actual experience what makes missions difficult or unsuccessful aids in tracking mission progress. Guiding progress toward mission accomplishment involves scheduling activities, tracking tasks and suspenses, alerting others when their support will be required, and making adjustments as required.

Adapting to Changes

9-26. Competent and realistic leaders also keep in mind that friction and uncertainty can and will always affect plans; generally, no plan survives initial contact with the enemy. The leader must therefore be prepared to replace portions of the original plan with new ideas and initiatives. Leaders must have the confidence and resilience to fight through setbacks, staying focused on the intent two levels up and the mission. Leaders preserve freedom of action by adapting to changing situations. They should be in a position to keep their people mission-focused, motivated, and able to react with agility to changes while influencing the team to accomplish the mission as envisioned in the plan.

9-27. Adjustments are needed when facing obstacles that were not anticipated. In increasingly busy times, leaders need to provide an environment in which subordinates can focus and accomplish critical tasks. Minimizing and preventing distractions allows subordinates to pay full attention to mission accomplishment. Leaders need to ensure that additional taskings are within the capabilities of the unit or organization. If they are not, the leader needs to seek relief by going to superiors and clarify the impact that the additional workload has on the unit. Experienced leaders anticipate cyclical workloads and schedule accordingly. Competent leaders will make good decisions about when to press Soldiers and civilians and when to ease back and narrow focus on the one or two most important tasks if performance is in decline.

9-28. Leaders constantly scan what is going on in the work environment and the mission. With this awareness of the situation, the leader will recognize when the situation has changed or when the plan is not achieving the desired effects. If the situation changes significantly, leaders will consider options for proceeding, including the review of any contingencies that were developed to deal with new circumstances. Leaders make on-the-spot adjustments in the course of action to keep moving toward designated goals.

Managing Resources

9-29. A main responsibility of leaders—whether officers, NCOs, or Army civilians—is to accomplish the assigned mission, which includes making the best use of available resources. Some Army leaders specialize in managing single categories of resources, such as ammunition, food, personnel, and finances, but all leaders have an interest in overseeing that all categories of resources are provided and used wisely by their teams.

9-30. Managing resources consists of multiple steps that require different approaches and even different skills. In many cases, Army leaders need to acquire needed resources for themselves or others. Resources can take the form of money, materiel, personnel, and time. The acquisition process can be a relatively straightforward process of putting in a request through proper channels. Other times a leader may need to be more creative and resourceful. In such cases, the effective use of influence tactics (see Chapter 7) will likely be instrumental in successfully acquiring needed resources.

9-31. After resources have been acquired, leaders are responsible for allocating them in an impartial manner that recognizes different needs and priorities. A leader may have multiple requests for limited resources and will need to make decisions about the best distribution of resources. Doing so in a way that recognizes and resolves potential ethical dilemmas requires a firm grounding in the Army Values (see Chapter 4). Ultimately, a leader must decide how to best allocate resources in ways to meet the Army's mission. Leaders need to deal openly and honestly with their allocation decisions and be prepared to handle reactions from those who may feel that their requests were not handled fairly or effectively.

9-32. Leaders should evaluate if the limited resources were used wisely and effectively. Do the resources advance the mission of the Army and the organization? Conversely, were the resources squandered or used in ways that did not enhance the effectiveness of the individual, unit, or the Army as a whole? In cases in which resources were not used wisely, a leader should follow this evaluation with appropriate counseling for those who are accountable for the resources in question.

ACCOMPLISHING MISSIONS

...[S]chools and their training offer better ways to do things, but only through experience are we able to capitalize on this learning. The process of profiting from mistakes becomes a milestone in learning to become a more efficient soldier.

William G. Bainbridge
Sergeant Major of the Army (1975-1979)

9-33. A critical element of getting results is adopting measures that support a capability for consistent accomplishment. Achieving consistent results hinges on doing all the right things addressed by the other competencies—having a clear vision, taking care of people, setting the right example, building up the organization, encouraging leader growth, and so on. Consistent performance can be achieved by using techniques to—

- Monitor collective performance.
- Reinforce good performance.
- Implement systems to improve performance.

MONITORING PERFORMANCE

9-34. The ability to assess a situation accurately and reliably against desired outcomes, established values, and ethical standards is a critical tool for leaders to achieve consistent results and mission success. Assessment occurs continually during planning, preparation, and execution; it is not solely an after-the-fact evaluation. Accurate assessment requires instinct and intuition based on experience and learning. It also demands a feel for the reliability and validity of information and its sources. Periodic assessment is necessary to determine organizational weaknesses and prevent mishaps. Accurately determining causes is essential to training management, developing subordinate leadership, and initiating quality improvements.

Assessment Techniques

9-35. There are many different ways to gather information for assessment purposes. These include asking team members questions to find out if information is getting to them, meeting people to inquire if tasks and objectives are appropriate, and checking for plan synchronization. Assessing can also involve researching and analyzing electronic databases. No matter which techniques the leader explores, it is important that information be verified as accurate.

9-36. Although staff and key subordinates manage and process information for organizational and strategic leaders, this does not relieve them from the responsibility of analyzing information as part of the decision-making process. Often, leaders draw information from various sources to be able to compare the information and create a multidimensional picture. Often, leaders accomplish this by sending out liaison officers thoroughly familiar with their commander's intent as their eyes and ears.

9-37. While personal presence and the eyes and ears deliver much useful information, leaders can also exploit technologies for the purpose of timely assessment. In the world of digital command and control, commanders can set various command and control systems to monitor the status of key units, selected enemy parameters, and critical planning and execution time lines. They may establish prompts in the information systems that warn of selected critical events. Information systems may provide alerts about low fuel levels in maneuver units, tight management time lines among aviation crews, or massing enemy artillery. Management information systems in institutional settings may track the amount of email or new documents created.

9-38. It is sometimes dangerous to be too analytical when dealing with automated information or limited amounts of time. When analyzing information, leaders should guard against rigidity, impatience, or overconfidence that may bias their analysis.

Designing an Effective Assessment System

9-39. The first step in designing an effective assessment system is to determine the purpose of the assessment. While purposes vary, most fall into one of the following categories:

- Evaluate progress toward organizational goals, such as using an emergency deployment readiness exercise to check unit readiness or monitoring progress of units through stages of reception, staging, onward movement, and integration.
- Evaluate the efficiency of a system: the ratio of the resources expended to the results gained, such as comparing the amount of time spent performing maintenance to the organization's readiness rate.
- Evaluate the effectiveness of a system: the quality of the results it produces, such as analyzing the variation in Bradley gunnery scores.
- Compare the relative efficiency or effectiveness against standards.
- Compare the behavior of individuals in a group with the prescribed standards, such as Army physical fitness test or gunnery scores.
- Evaluate systems supporting the organization, such as following up "no pay dues" to see what the NCO support channel did about them.

9-40. While systems and leader proxies can greatly assist in assessing organizational performance, the leader remains central by spot-checking people, performance, equipment, and resources. Leaders adopt best business practices, use performance indicators to check things, and ensure the organization meets standards while moving toward the goals the leader has established.

9-41. While assessing, good leaders find opportunities to engage in impromptu coaching. Junior leaders can learn spot-checking by watching experienced first sergeants or command sergeants major observe daily training or conduct uniform inspections. Pay attention to how these experienced leaders' eyes sweep across Soldiers, weapons, and equipment and note discrepancies and successes. It demonstrates how experience makes supervising, inspecting, and correcting becomes a routine part of daily duties.

REINFORCING GOOD PERFORMANCE

9-42. To accomplish missions consistently leaders need to maintain motivation among the team. One of the best ways to do this is to recognize and reward good performance. Leaders who recognize individual and team accomplishments will shape positive motivation and actions for the future. Recognizing individuals and teams in front of superiors and others gives those contributors an increased sense of worth. Soldiers and civilians who feel their contributions are valued are encouraged to sustain and improve performance.

9-43. Leaders should not overlook giving credit to subordinates. Sharing credit has enormous payoffs in terms of building trust and motivation for future actions. A leader who understands how individuals feel about team accomplishments will have a better basis for motivating individuals based on their interests.

IMPROVING ORGANIZATIONAL PERFORMANCE

9-44. High performing units are learning organizations that take advantage of opportunities to improve performance. Leaders need to encourage a performance improvement mindset that allows for conformity but goes beyond meeting standards to strive for increased efficiencies and effectiveness. Several actions are characteristic of performance improvement:

- Ask incisive questions about how tasks can be performed better.
- Anticipate the need for change and action.
- Analyze activities to determine how desired end states are achieved or affected.
- Identify ways to improve unit or organizational procedures.
- Consider how information and communication technologies can improve effectiveness.
- Model critical and creative thinking and encourage it from others.

9-45. Too often, leaders unknowingly discourage ideas. As a result, subordinates become less inclined to approach leaders with new ideas for doing business. From their viewpoint leaders respond to subordinates' ideas with reactions about what is and is not desired. This can be perceived as closed-mindedness and under-appreciation of the Soldier's or civilian's insight. "We've tried that before." "There's no budget for that." "You've misunderstood my request." "Don't rock the boat." These phrases can kill initiative and discourage others from even thinking about changes to improve the organization. Leaders need to encourage a climate of reflection about the organization and encourage ideas for improvement. The concept of lifelong learning applies equally to the collective organization as well as to the individual.

COMPETENCIES APPLIED FOR SUCCESS

The American people expect only one thing from us: That we will win. What you have done is no more than they expect. You have won.

General Gordon Sullivan
Vice Chief of Staff of the Army
Addressing the Third Army staff following the Operation Desert Storm victory (1991)

9-46. Army history has many examples of units succeeding in accomplishing their mission consistently and ethically because of competent, multiskilled leaders. Achieving results consistently and ethically does not merely pertain to combat or military leadership. Competent military and civilian leaders pursue excellence wherever and whenever possible.

9-47. Competent leaders ensure that all organization members know the important roles they play every day. They look for everyday examples occurring under ordinary circumstances: how a Soldier digs a fighting position, prepares for guard duty, fixes a radio, or lays an artillery battery; or how an Army civilian improves maintenance procedures, processes critical combat supplies, and supports the families of deploying service members. Competent leaders know each of these people is contributing in an important way to the Army mission. They appreciate the fact that to accomplish the Army's mission with consistency and ethics, it requires a collection of countless teams, performing countless small tasks to standard every day.

9-48. Competent leaders are also realists. They understand that excellence in leadership does not mean perfection. On the contrary, competent leaders allow subordinates room to learn from their mistakes as well as their successes. In an open and positive work climate, people excel to improve and accept calculated risks to learn. It is the best way to improve the force and the only way to develop confident leaders for the future. Competent and confident leaders tolerate honest mistakes that do not result from negligence, because achieving organizational excellence is not a game to reach perfection. It involves trying, learning, trying again, and getting better each time. However, even the best efforts and good intentions cannot take away an individual's responsibility for their own actions.

9-49. At the end of the day or a career, Soldier and Army civilian leaders can look back confidently that their efforts have created an Army of consistent excellence. Whether they commanded an invasion force of thousands or supervised a technical support section of three people, they made a positive difference.

Achieving Success and Leadership Excellence

GEN Matthew B. Ridgway successfully led the 82d Airborne Division and XVIII Airborne Corps during World War II. He later commanded the Eighth (U.S.) Army during the Korean War. GEN Ridgway exemplified the qualities of the competent and multiskilled Army leader. His knowledge of American Soldiers, other Services, allies, foreign cultures, and the overall strategic situation led him to certain expectations. Those expectations gave him a baseline from which to assess his command once he arrived in theater. He continually visited units throughout the Eighth Army area, talked with Soldiers and their commanders, assessed command climate, and took action to mold attitudes with clear intent, supreme confidence, and unyielding tactical discipline.

GEN Ridgway constantly sought to develop and mentor subordinate commanders and their staffs by sharing his thoughts and expectations of combat leadership. He frequently visited the frontlines to feel the pulse of the fighting forces, shared their hardships, and demanded they be taken care of. He took care of his troops by pushing the logistic systems to provide creature comforts as well as war supplies. He eliminated the skepticism of purpose, gave Soldiers cause to fight, and helped them gain confidence by winning small victories. GEN Ridgway led by example.

His actions during four months in command of the Eighth Army prior to his appointment as United Nations Supreme Commander bring to life the leader's competencies. He left a legacy that leaders can operate within the spheres of all levels of leadership to accomplish their mission consistently and ethically.

Chapter 10
Influences on Leadership

10-1. Each day as a leader brings new challenges. Some of these challenges are predictable based on experiences. Some are unpredictable, surfacing because of a situation or place in time in which Soldiers find themselves. Leaders must be prepared to face the effects of stress, fear in combat, external influences from the media, the geopolitical climate, and the impact of changing technology.

10-2. Some of these factors are mitigated through awareness, proper training, and open and frank discussion. The Army must consider these external influences and plan accordingly. An effective leader recognizes the tools needed to adapt in changing situations. (See paragraph 10-48 for further discussion.)

CHALLENGES OF THE OPERATING ENVIRONMENT

The role of leadership is to turn challenges into opportunities.

General Dennis J. Reimer
Chief of Staff, Army (1995-1999)

ADAPTING TO EVOLVING THREATS

10-3. America's Army of the 21st century must adapt to constantly evolving threats while taking advantage of the latest technological innovations and adjusting to societal changes. As part of the United States Armed Forces, the Army is guided by a broader National Military Strategy outlining how to—

- Protect the United States.
- Prevent conflict and surprise attacks.
- Prevail against adversaries threatening our homeland, deployed forces, or allies and friends.

10-4. The National Military Strategy also sets priorities for success and changes with each administration and addresses new challenges our country faces. The uncertain nature of the threat will always have major impact on Army leadership. For the Army, a new era began in 1989 with the fall of the Berlin Wall and the subsequent collapse of the Soviet Union. Since 11 September 2001, the War on Terrorism has become America's main effort and long-term security focus. In addition to adapting to evolving issues, U.S. forces must also remain capable to conduct full spectrum operations. This mandates that the Army, as an essential component of America's war effort, be fully capable of seamless shifts across the spectrum of conflict. This blurring of the lines between war and peace make the challenges that leaders face constant and unpredictable.

10-5. Agility and adaptability at all leadership levels of Army organizations are becoming more important to address situations that cannot be fully anticipated. In the new operational environment, the importance of direct leaders— noncommissioned officers and junior officers—making the right decisions in stressful situations has taken on a new significance. Decisions and actions taken by direct-level leaders—the sergeants and lieutenants carrying out the missions—can easily have major strategic-level and political implications.

10-6. U.S. forces in Afghanistan and Iraq have experienced many situations requiring a balanced application of tactical and diplomatic measures. In most of these tactical confrontations, junior leaders ensure mission accomplishment by reacting appropriately and within the bounds of their commanders' intent.

The Influence of the Media

10-7. Another influence on leadership is the media. The media can be both an asset and impediment to the leader. Embedded media, like those during Operation Iraqi Freedom, can tell the story from the Soldier's perspective to an anxious Nation back home. The media can provide real-time information, sometimes unfiltered and raw, which the enemy could exploit as a means to change the regional political climate.

10-8. Leaders must ensure subordinate leaders and Soldiers are trained to deal with the media and understand the long-term effects of specific stories and images. The morale of those serving and the Nation may be affected if the overall view presented by the media is overly negative or that military actions are in vain. These can adversely impact recruiting, retention, and the treatment of veterans for years to follow. Leaders can counter-act negatives by using media opportunities to explain how the Army mission serves national interests and how Soldiers dedicate themselves to accomplishing the mission.

Multicomponent and Joint Environment

10-9. Soldiers find themselves serving with members of other Services, the Reserve Components, and other countries' forces more than ever before. Understanding the unique cultures and subcultures of these various groups can be essential to success in a volatile and changing world.

10-10. Leaders must be aware that while most of the policies and regulations for Soldiers apply across the board, specific differences apply in the promotion, pay, benefit, and retirement systems of the Reserve Components. Knowledge of the differences is essential for effectively employing all components.

10-11. Within the Army, leaders should recognize the existence of subcultures such as the special operations, law enforcement, medical, and branch-specific communities. Members of these subcultures cross components and Services during their careers for specific assignments. Consequently, leaders involved in conducting operations need to understand how members of these specialized units train and work. Often, they approach missions from a different perspective and sometimes use unconventional methods to accomplish them. Special operations forces usually operate in small, independent teams and frequently interface with local civilians and members of other governmental agencies. For operational reasons, they may not be required to disclose routine information about their units like conventional forces. Logisticians and operations planners may need innovative solutions to provide special operations forces autonomy while allowing the joint task force or other commanders to maintain visibility and control over these assets and to provide the special operators the logistical support necessary.

10-12. Other subcultures, such as law enforcement, follow norms established by their branches and share experiences developed through specific assignments and schools. These functional subcultures can be useful as a means to exchange knowledge and provide corporate solutions when the Army needs answers from subject matter experts.

The Geopolitical Situation

10-13. Though the world continues to become more connected by technology and economic growth, it remains very diverse and divided by religions, cultures, living conditions, education, and health. Within the political sphere of influence, maintaining our presence in foreign countries through a careful mix of diplomatic and military arrangements remains an important challenge. Leaders must be aware that the balance between diplomacy and military power is fragile. Army leaders must consistently consider the impact on local civilians, as well as on cultural and religious treasures, prior to committing firepower.

10-14. Tomorrow's leaders will be expected to operate in many different environments worldwide. While most Soldiers speak English as their first language, continued deployments and global interaction will require an understanding of other languages and cultures. Forecasts predict the Chinese, Hindu, Arabic, and Spanish languages will gain speakers in the years to come. Leaders will need to become multilingual and study the cultures and histories of other regions of interest. A vehicle for gaining this knowledge of the geopolitical situation is technology.

CHANGING WITH TECHNOLOGY

10-15. While the stresses of combat have been constant for centuries, another aspect of the human dimension has assumed increasing importance: the effect of rapid technological advances on organizations and people. Although military leaders have always dealt with the effect of technological changes, these changes are different from before. It is forcing the Army and its leaders to rethink and redesign itself.

10-16. Modern Army leaders must stay abreast of technological advances and learn about their applications, advantages, and requirements. Together with technical specialists, leaders can make technology work for the warrior. The right technology, properly integrated, will increase operational effectiveness, battlefield survivability, and lethality.

10-17. Technological challenges facing the Army leadership include—

- Learning the strengths and vulnerabilities of different technologies that support the team and its mission.
- Thinking through how the organization will operate with other organizations that are less or more technologically complex, such as operating with joint, inter-Service, and multinational forces.
- Considering the effect of technology on the time available to analyze problems, make a decision, and act. Events happen faster today, and the stress encountered as an Army leader is correspondingly greater.
- Leveraging technology to influence virtual teams given the increasing availability and necessity to use reach-back and split-based operations.

Virtual team refers to any team whose interactions are mediated by time, distance, or technology.

10-18. Technology can also lead to operational issues. A growing reliance on the new global positioning system (GPS) navigation technology since the Desert Storm era decreased emphasis on manual land navigation skills in training, thus rendering forces more vulnerable if the technology fails or is wrongly programmed. Part of the leadership challenge became to determine how to exploit GPS technology while guarding against its weaknesses. The answer was improved training. It included reintroducing essential back-up land navigation training, emphasizing the availability of adequate battery supplies, and detailed instructions on the maintenance and operation of the GPS receiver equipment.

10-19. Leaders not on-site with the Soldiers must not discount the fear the Soldiers may be experiencing. A leader who does not share the same risks could easily fall into the trap of making a decision that could prove unworkable given the psychological state of the Soldiers. Army leaders with command and control over a distributed or virtual team should ask for detailed input from the Soldiers or subordinate commanders who are closer to the action and can provide the most accurate information about the situation.

10-20. Technology is changing the leadership environment in many aspects, especially the amount of information available for decision makers. Although advances in electronic data processing allow the modern leader to handle large amounts of information easier than ever before, a possible second-order effect of enhanced technology is information overload.

10-21. Too much information is as bad as not enough. Leaders must be able to sift through the information provided to them, analyze and synthesize it, and forward only the important data up the chain of command. Senior leaders rely on their subordinates to process information for them, isolating critical information to expedite decisions. Leaders owe it to their subordinates to design information gathering and reporting procedures that do not create more work for already stretched staffs and units.

10-22. Army leaders and staffs have always needed to determine mission-critical information, prioritize incoming reports, and process them quickly. The volume of information provided by current technology makes this ability even more critical. The answer lies in the agile and adaptable human mind. Sometimes a nontechnological approach can divert the flood of technological help into channels the leader and staff can capably manage. For example, a well-understood commander's intent and thought-through commander's critical information requirements will help free leaders from nonessential information overload. The Army concept of mission command is even more important in an environment of information overload. Mission command delegates most decisions to lower echelons in order to free higher echelons for critical decisions only they can make. Army leaders should continue to resist the lure of centralized decision making even though they have more information available to them than ever before.

SYNCHRONIZING SYSTEMS

10-23. Today's Army leaders require systems understanding and more technical and tactical knowledge than ever before. Leaders must be aware of the fine line between a healthy questioning of new systems' capabilities and an unreasonable hostility that rejects the advantages technology offers. The adaptable leader remains aware of the capabilities and shortcomings of advanced technology and ensures subordinates do as well.

10-24. All leaders must consider systems in their organization—how they work together, how using one affects the others, and how to get the best performance from the whole. They must think beyond their own organizations and consider how the actions of their organization can influence other organizations and the team as a whole.

10-25. Technology is also changing battlefield dispersal and the speed of operations. Instant global communications are accelerating the pace of military actions. GPS and night vision capabilities mean the Army can fight at night and during periods of limited visibility—conditions that used to slow things down. Additionally, nonlinear operations make it more difficult for commanders to determine critical points on the battlefield. (FM 3-0 discusses continuous operations.)

10-26. Modern technology has also increased the number and complexity of skills the Army requires. Army leaders must carefully manage low-density occupational specialties and ensure critical positions are filled with properly trained people who maintain proficiency in these perishable high-tech skills. Army leaders must balance leadership, personnel management, and training management to ensure their organizations are assigned people with the appropriate specialty training and that the entire organization stays continuously trained, certified, and ready.

STRESS IN COMBAT

All men are frightened. The more intelligent they are, the more they are frightened. The courageous man is the man who forces himself, in spite of his fear, to carry on.

General George S. Patton, Jr.
War As I Knew It (1947)

10-27. Combat is sudden, intense, and life threatening. It is the Soldier's job to kill in combat. Unfortunately, combat operations may involve the accidental killing of innocent men, women, and children. Soldiers are unsure how they will perform in combat until that moment comes. The stresses experienced in combat and even the stress preparing for, waiting for, and supporting combat can be substantial.

10-28. Leaders must understand this human dimension and anticipate Soldiers' reactions to stress. It takes mental discipline and resilience to overcome the plan going wrong, Soldiers becoming wounded or dying, and the enemy attacking unexpectedly.

10-29. When preparing for war, leaders must thoroughly condition their Soldiers to deal with combat stress during all phases of operations—mobilization, deployment, sustainment, and redeployment. (See FM 6-22.5 for more on combat stress and FM 3-0 for descriptions of specific deployment phases.) The most potent countermeasures to confront combat stress and to reduce psychological breakdown in combat are—

- Admit that fear exists when in combat.
- Ensure communication lines are open between leaders and subordinates.
- Do not assume unnecessary risks.

- Provide good, caring leadership.
- Treat combat stress reactions as combat injuries.
- Recognize the limits of a Soldier's endurance.
- Openly discuss moral implications of behavior in combat.
- Reward and recognize Soldiers and their families for personal sacrifices.

10-30. Units are stabilized during mobilization and in preparation for deployment. Stabilization allows leaders and Soldiers to build a trust relationship while the unit undergoes rigorous combat skills certification or theater-specific training. Confidence in leaders, comrades, training, and equipment are key factors for combat success.

10-31. During initial deployment, units should be eased into the mission. A daytime operation could precede a night raid, for example. Training and drill can continue while leaders deepen a personable leader-to-led relationship with their Soldiers based on trust and not fear of rank and duty position.

10-32. During sustaining operations, units at all levels should discuss and absorb critical operations experiences and help individuals cope with initial combat stress. Soldiers can be encouraged to reveal their true feelings within their circle of warrior comrades. If the unit suffered casualties, leaders should openly discuss their status. In this phase, it is important to keep people informed about wounded and evacuated team members and to weigh the unit's losses and successes. Memorial services should be held to honor the fallen. Soldiers and leaders who do not succeed during operations should be retrained, counseled, or reassigned. The unit should be allocated appropriate rest periods between missions. Ensure Soldiers with serious issues have access to mental health professionals if necessary.

10-33. When preparing to redeploy, Soldiers should talk about their experiences. Leaders and commanders should be available first and refer or bring in backup like psychologists or chaplains when needed. During this phase, leaders must emphasize that Soldiers have an obligation to remain disciplined, just as they were during deployment. Soldiers must participate in provided reintegration screening and counseling. Leaders should stress that it is acceptable, and not shameful, to seek appropriate psychological help.

10-34. Once returned to their home station, organizations and units generally remain stabilized to further share common experiences before the individuals are released to new assignments. This can be difficult for returning Reserve Component forces that are often released very soon after redeployment.

10-35. When possible, Soldiers should have unfettered access to medical experts and chaplains to continue their physical and psychological recovery. Experts helping and treating the psychologically wounded must work hand-in-hand with the unit chain of command to stress the importance of maintaining good order and discipline. Aggressive or criminal behavior to compensate for wartime experiences is not tolerated.

10-36. The Army has implemented a comprehensive mental health recovery plan for all returning Soldiers to counter post-traumatic stress disorder. Sound leadership, unit cohesion, and close camaraderie are essential to assure expeditious psychological recovery from combat experiences.

Overcoming Fear in Battle

Sure I was scared, but under the circumstances, I'd have been crazy not to be scared....There's nothing wrong with fear. Without fear, you can't have acts of courage.

Sergeant Theresa Kristek
Operation Just Cause, Panama (1989)

10-37. Leaders need to understand that danger and fear will always be a part of their job. Battling the effects of fear does not mean denying them. It means recognizing fear and effectively dealing with it. Fear is overcome by understanding the situation and acting with foresight and purpose to overcome it. Army leaders must expect fear to take hold when setbacks occur, the unit fails to complete a mission, or there are casualties. Fear can paralyze a Soldier. Strong leaders share the same risks with their Soldiers, but use competence and extensive training to gain their Soldiers' trust and loyalty. The sights and sounds of the modern battlefield are terrifying. So is fear of the unknown. Soldiers who see their friends killed or wounded suddenly have a greater burden—they become aware of their own mortality.

10-38. Combat leadership is a different type of leadership where leaders must know their profession, their Soldiers, and the tools of war. Direct leaders have to be strong tacticians and be able to make decisions and motivate Soldiers under horrific conditions. They must be able to execute critical warrior tasks and drills amidst noise, dust, explosions, confusion, and screams of the wounded and dying. They have to know how to motivate their Soldiers in the face of adversity.

10-39. One leader who exhibited all these traits and helped his men overcome the fear of battle was Lieutenant Rick Rescorla.

A Fearless Leader—Twice a Hero

One of the "young Soldiers" who fought with LTC Harold Moore at the well-known battle of Ia Drang, Vietnam, in late 1965 was LT Cyril Richard "Rick" Rescorla. He was British, the epitome of a warrior, already battle-hardened by time spent in Cyprus and Rhodesia at the age of 24. Rescorla came to America to join the fight in Vietnam.

LTC Moore called him the best platoon leader he ever saw. His troops loved him for his spirit and fearlessness. The night after an entire company of the 2nd Battalion, 7th Cavalry was virtually annihilated at Landing Zone X-Ray, Rescorla's company was ordered to replace them on the perimeter at the foot of the Chu Pong ridge.

That night, the young lieutenant did all the right things to prepare his Soldiers for battle: studied the terrain, relocated foxholes, laid booby traps, and repositioned weapons. The best thing he did was display confidence.

Sometime after midnight, he started singing a slow Cornish mining tune: "Going Up Cambourne Hill Coming Down." One of his sergeants remembers Rescorla stopping by his foxhole to check on him and analyze his fields of fire.

"We all thought we were going to die that night," the sergeant said, "and he gave us our courage back. I figured if he's walking around singing, the least I can do is stop trembling."

The next morning Bravo Company defended against four assaults, killing over 200 enemy soldiers while sustaining only a few injuries. However, their task was not done. The next day when the battalion marched into a vicious ambush, Rescorla's men were called on to rescue them. Once again, the lieutenant arrived under fire and immediately lifted the spirits of weary Soldiers who thought they were finished.

Rescorla left Vietnam and returned to civilian life. He finished out his career in the Army Reserve, achieving the rank of colonel. He was the vice president for corporate security at Morgan Stanley Dean Witter & Company on 11 September 2001, the day a jet plowed into the World Trade Center.

Once again, Rescorla was cool under pressure. His military leadership experience served him well as he led his company's nearly 2,700 employees to safety. As the employees left the building, Rescorla remained inside searching for stragglers, determined to leave no one behind. He was last seen near the stairwell of the tenth floor, reassuring everyone that they would be all right. It is rumored that he sang his Cornish song again and led everyone in renditions of "God Bless America."

Rescorla called his wife and told her she had made his life. One of his last phone calls before he died was to an old friend from Vietnam, Dan Hill.

"Typical Rescorla," Hill recalled. "Incredible under fire."

10-40. What carries Soldiers through the terrible challenges of combat and operating in support under hazardous conditions is good preparation, planning, and rigorous training. Realistic training developed around critical tasks and battle drills is a primary source for the resilience and confidence to win along with the ability to gut it out when things get tough, even when things look hopeless. It is leader competence, confidence, agility, courage, and resilience that help units persevere and find workable solutions to the toughest problems. The Warrior Ethos and resilience mobilize the ability to forge victory out of the chaos of battle to overcome fear, hunger, deprivation, and fatigue and to accomplish the mission no matter what the odds.

THE WARRIOR MINDSET

10-41. It is important for Soldiers to acquire and maintain a warrior mindset when serving in harm's way. Resilience and the Warrior Ethos apply in more situations than those requiring physical courage. Sometimes leaders will have to carry on for long periods in very difficult situations. The difficulties Soldiers face may not only be ones of physical danger, but of great physical, emotional, and mental strain.

10-42. An essential part of the warrior mindset is discipline. Discipline holds a team together, while resilience, the Warrior Ethos, competence, and confidence motivate Soldiers to continue the mission against all odds. Raw physical courage causes Soldiers to charge a machine gun but resilience, discipline, and confidence backed by professional competence help them fight on when they are hopelessly outnumbered and living under appalling conditions.

STRESS IN TRAINING

War makes extremely heavy demands on the soldier's strength and nerves. For this reason, make heavy demands on your men in peacetime.

Field Marshall Erwin Rommel
Infantry Attacks (1937)

10-43. As Erwin Rommel wrote in 1937, it is still valid for the complex combat environment of the War on Terrorism: Training to high standards—using scenarios that closely resemble the stresses and effects of the real battlefield—is essential to victory and survival in combat.

10-44. Merely creating a situation for subordinates and having them react does not induce the kind of stress required for combat training. A meaningful and productive mission with detailed constraints and limitations and high standards of performance induces a basic level of stress. To reach a higher level of reality, leaders must add unanticipated conditions to the basic stress levels of training to create a demanding learning environment.

DEALING WITH THE STRESS OF CHANGE

10-45. Since the end of the Cold War, the Army has gone through many changes—dramatic decreases in the number of Soldiers and Army civilians in all components, changes in assignment policies, base closings, new organizational structures, and a host of other shifts that put stress on Soldiers, Army civilians, and their families. Despite the Army's reduced personnel strength, deployments to conduct stability operations and to fight the spread of terrorism have increased considerably. While adapting to the changes, Army leaders continuously have to sustain the force and prepare the Soldiers of all components for the stresses of combat.

10-46. To succeed in an environment of continuous change, Army leaders emphasize the constants of the Army Values, teamwork, and discipline while helping their people anticipate change, adapt to change, and seek new ways to improve. Competent leadership implies managing change, adapting, and making it work for the entire team. Leaders determine what requires change. Often, it is better to build on what already exists to limit stress.

10-47. Stress will be a major part of the leadership environment, both in peace and war. Major sources of stress include an ever-changing geopolitical situation, combat stress and related fears, the rapid pace of change, and the increasing complexity of technology. A leader's character and professional competence are important factors in mitigating stress for the organization and achieving mission accomplishment, despite environmental pressures and changes. When dealing with these factors, adaptability is essential to success.

TOOLS FOR ADAPTABILITY

10-48. Adaptability is an individual's ability to recognize changes in the environment, identify the critical elements of the new situation, and trigger changes accordingly to meet new requirements.

Adaptability is an effective change in behavior in response to an altered situation.

10-49. Adaptable leaders scan the environment, derive the key characteristics of the situation, and are aware of what it will take to perform in the changed environment. Leaders must be particularly observant for evidence that the environment has changed in unexpected ways. They recognize that they face highly adaptive adversaries, and operate within dynamic, ever-changing environments. Sometimes what happens in the same environment changes suddenly and unexpectedly from a calm, relatively safe operation to a direct fire situation. Other times environments differ (from a combat deployment to a humanitarian one) and adaptation is required for mind-sets and instincts to change.

10-50. Highly adaptable leaders are comfortable entering unfamiliar environments. They have the proper frame of mind for operating under mission command orders in any organization (see FM 6-0). Successful mission command results from subordinate leaders at all echelons exercising disciplined initiative within the higher commander's intent. All adaptable leaders can quickly assess the situation and determine the skills needed to deal with it. If the skills they learned in the past are not sufficient for success in the new environment, adaptable leaders seek to apply new or modified skills and applicable competencies.

10-51. Adaptive leadership includes being an agent of change. This means helping other members of the organization, especially key leaders, to recognize that an environment is changing and building consensus as change is occurring. As this consensus is built, adaptive leaders can work to influence the course of the organization. Depending on the immediacy of the problem, adaptive leaders may use several different methods for influencing their organization. These can range from "crisis action meetings" (when time is very short) to publishing white papers or other "thought pieces" that convey the need for change (when more time is available).

10-52. Leaders lacking adaptability enter all situations in the same manner and often expect their experience in one job to carry them to the next. Consequently, they may use ill-fitting or outdated strategies. Failure to adapt may result in poor performance in the new environment or outright organizational failure.

10-53. Deciding when to adapt is as important as determining how to adapt. Adaptation does not produce certainty that change will improve results. Sometimes, persistence on a given course of action may have merit over change.

> *Nothing in the world can take the place of persistence. Talent will not…. Genius will not …. Education will not …. Persistence and determination alone are omnipotent. The slogan "press on" has solved and always will solve the problems of the human race.*
>
> Calvin Coolidge
> President of the United States (1923-1929)

10-54. Adaptable leaders are comfortable with ambiguity. They are flexible and innovative—ready to face the challenges at hand with the resources available. The adaptable leader is most likely a passionate learner, able to handle multiple demands, shifting priorities and rapid change smoothly. Adaptable leaders see each change thrust upon them as an opportunity rather than a liability.

10-55. Adaptability has two key components:

- The ability of a leader to identify the essential elements critical for performance in each new situation.
- The ability of a leader to change his practices or his unit by quickly capitalizing on strengths and minimizing weaknesses.

10-56. Like self-awareness, adaptability takes effort. To become adaptable, leaders must challenge their previously held ideas and assumptions by seeking out situations that are novel and unfamiliar. Leaders who remain safely inside their comfort zone provided by their current level of education, training, and experience will never learn to recognize change or understand the inevitable changes in their environment. Adaptability is encouraged by a collection of thought habits. These include open-mindedness, ability to consider multiple perspectives, not jumping to conclusions about what a situation is or what it means, willingness to take risks, and being resilient to setbacks. To become more adaptable, leaders should—

- **Learn to adapt by adapting.** Leaders must go beyond what they are comfortable with and must get used to experiencing the unfamiliar through diverse and dynamic challenges. For example, the Army's best training uses thinking like an enemy to help leaders recognize and accept that no plan survives contact with the enemy. This encourages adaptive thinking. Adaptive training involves variety, particularly in training that may have become routine.
- **Lead across cultures.** Leaders must actively seek out diverse relationships and situations. Today's joint, interagency, and multinational assignments offer challenging opportunities to interact across cultures and gain insight into people who think and act differently than most Soldiers or average U.S. citizens. Leaders can grow in their capacity for adaptability by seizing such opportunities rather than avoiding them.
- **Seek challenges.** Leaders must seek out and engage in assignments that involve major changes in the operational environment. Leaders can be specialists, but their base of experience should still be broad. As the breadth of experience accumulates, so does the capacity to adapt. Leaders who are exposed to change and embrace new challenges will learn the value of adaptation. They carry forward the skills to develop adaptable Soldiers, civilians, units, and organizations.

10-57. While adaptability is an important tool, leaders at all levels must leverage their cognitive abilities to counteract the challenges of the operational environment through logical problem solving processes. FM 5-0 discusses these in detail.

PART FOUR

Leading at Organizational and Strategic Levels

All professional Army leaders consistently prepare themselves for greater responsibilities while mastering core leader competencies. By the time they become organizational and strategic leaders, they should be multiskilled leaders who can comfortably operate at all levels of leadership and apply their vast experiences and knowledge for success across the spectrum of conflicts. They oversee continuous transformation of the Army and respond to evolving operational environments. They also mentor and develop the leadership of the future force.

Chapter 11

Organizational Leadership

11-1. Whether they fight for key terrain in combat or work to achieve readiness in peacetime training, organizational leaders must be able to translate complex concepts into understandable operational and tactical plans and decisive action. Organizational leaders develop the programs and plans, and synchronize the appropriate systems allowing Soldiers in small units to turn tactical and operational models into action.

11-2. Through leadership by example, a wide range of knowledge, and the application of leader competencies, organizational leaders build teams of teams with discipline, cohesion, trust, and proficiency. They focus their organizations down to the lowest level on the mission ahead by disseminating a clear intent, sound operational concepts, and a systematic approach to execution.

LEADING

11-3. Successful organizational leadership tends to build on direct leader experiences. Because they lead complex organizations, such as task forces, brigade combat teams, divisions, and corps, organizational leaders often apply elements of direct, organizational, and strategic leadership simultaneously. Highly accelerated operating tempos, compressed training cycles, contingency operations, and continual deployment cycles mandate leader agility. The modern organizational level leader must carefully extend his influence beyond the traditional chain of command by balancing his role of warrior with that of a diplomat in uniform.

LEADS OTHERS

The American soldier demonstrated that, properly equipped, trained and led, he has no superior among all of the armies in the world.

Lt. General Lucian K Truscott
Commanding General, 5th Army, World War II

11-4. Modern organizational leaders are multiskilled, multipurpose leaders. They have developed a strong background in doctrine, tactics, techniques, and procedures, as well as an appreciation for the geopolitical consequences of their application. From their personal experience at the operational and tactical levels, they have grown the instincts, intuition, and knowledge that form the understanding of the interrelation of tactical and operational processes (FM 3-0). Their refined tactical skills allow them to understand, integrate, and synchronize the activities of multiple systems, bringing all resources and systems to bear across the spectrum of conflicts.

11-5. Given the increased size of their organizations, organizational leaders influence more often indirectly than in person. They rely more heavily on developing subordinates and empowering them to execute their assigned responsibilities and missions. They should be able to visualize the larger impact on the organization and mission when making decisions. Soldiers and subordinate leaders, in turn, look to their organizational leaders to set achievable standards, to provide clear intent, and to provide the necessary resources.

11-6. Decisions and actions by organizational leaders have far greater consequences for more people over a longer time than those of direct leaders. Because the connections between action and effect are sometimes more remote and difficult to see, organizational leaders spend more time than direct leaders thinking and reflecting about what they are doing and how they are doing it. Organizational leaders develop clear concepts for operations as well as policies and procedures to control and monitor their execution.

Extends Influence Beyond the Chain of Command

11-7. While organizational leaders primarily exert direct influence through their chain of command and staff, they extend influence beyond their chain of command and organization by other means. These include persuasion, empowerment, motivation, negotiation, conflict resolution, bargaining, advocacy, and diplomacy. They often apply various skills when serving as military negotiators, consensus builders, and operational diplomats in joint, interagency, and multinational assignments. Chiefs of special directorates within and outside the Army also need these skills. As leaders, they affect the operational situation in their area of operations by extending influence through local leaders such as police chiefs, mayors, and tribal elders. Numerous experiences during Operation Iraqi Freedom have shown that the organizational leader, when effectively balancing the functions of combat leader and military diplomat, can set the stage for military, political, and social stability in assigned areas.

Leveraging Joint, Interagency, and Multinational Capabilities

11-8. Brigade combat teams, task forces, and battalions often participate in joint and multinational operations. Consequently, organizational leaders and their staffs must understand joint procedures and concerns, just as much as Army procedures and concerns. Additionally, corps or divisions may control forces of other nations. This means that corps, division, and even brigade combat team headquarters and below may have liaison officers from other nations. In some cases, U.S. staffs may have members of other nations permanently assigned, creating a multinational staff.

11-9. Today's operations present all Army leaders, particularly organizational leaders, with a nonlinear, dynamic environment. These varied conditions create an information-intense environment, challenging leaders to synchronize their efforts beyond the traditional military chain. Today's mission complexities might demand the full integration and cooperation of nonmilitary and nongovernmental agencies to accomplish missions.

Negotiating, Building Consensus and Resolving Conflicts

11-10. Leaders often must leverage negotiating skills to obtain the cooperation and support necessary to accomplish a mission beyond the traditional chain of command. During complex operations, different joint, interagency, and multinational contingents might operate under specific restraints by their national or organizational chains. This can result in important negotiations and conflict resolution versus a simpler process of merely issuing binding orders.

11-11. Successful negotiating involves communicating a clear position on relevant issues and integrating understanding of motives while conveying a willingness to bargain on other issues. This requires recognizing what is acceptable to the negotiating parties and achieving a workable compromise. Good negotiators visualize several possible end states while maintaining a clear idea of the optimal end state from the parent command's perspective.

11-12. In joint and multinational operations, leaders often have to create consensus by carefully persuading others about the validity of the U.S. position. They must convince others that the United States fully understands and respects their interests and concerns. The art of persuasion is an important method of extending influence. Working through controversy in a positive and open way helps overcome resistance to an idea or plan and build support. Proactively involving partners frees communications with them and places value on their opinions. Openness to discussing one's position and a positive attitude toward a dissenting view often diffuses conflict, increases mutual trust, and saves time.

LEADS BY EXAMPLE

If you are the leader, your people expect you to create their future. They look into your eyes, and they expect to see strength and vision. To be successful, you must inspire and motivate those who are following you. When they look into your eyes, they must see that you are with them.

General Gordon R. Sullivan
Hope is Not a Method (1996)

11-13. Army operations since the Cold War ended have shown all organizations must be capable of adapting to rapidly changing situations. It is often the ability to make quality decisions quickly and execute them within the enemy's decision cycle that determines who wins a sudden engagement or battle.

11-14. The Army's organizational leaders play a critical part when it comes to maintaining focus on fighting the enemy and not the plan. They are at the forefront of adapting to changes in the operational environment and exploiting emerging opportunities by applying a combination of intuition, analytical problem solving, systems integration, and leadership by example—as close to the action as feasible.

11-15. To see and feel the situation at hand and exert leadership by personal presence and example, organizational leaders position themselves as closely as possible to the front with all necessary means to maintain contact with critical combat elements and headquarters. The V Corps forward headquarters used by Lieutenant General William S. Wallace during Operation Iraqi Freedom was a compact, mobile command and control element that facilitated the effective leading of many complex organizations from the front. It consisted of approximately eighty key personnel, three command and control vehicles, and ten support vehicles. The general's forward headquarters was sufficiently mobile to enable him and key staff members to see and feel the battlespace and to maintain close contact with organizations in critical fights.

11-16. Proximity to the front provides today's organizational commanders with the required awareness to apply quick creative thinking in collaboration with subordinate leaders. It facilitates adjustments for deficiencies in planning and shortens reaction time when applying sound tactical and operational solutions to changing battlefield realities. In some areas of operation during Operation Iraqi Freedom, creative organizational leadership was instrumental in achieving a swift transition from mechanized warfare to stability operations centering on urban areas. Transition efforts mandated creating ad-hoc organizations, integrating new equipment and technologies, as well as adjusting the rules of engagement.

11-17. Organizational leaders represent the critical link to collecting, recording, and exploring the tactical and operational lessons learned. They ultimately direct the integration of critical experiences and new concepts into doctrine and future training. They leverage Army schools and combat training centers to coach and mentor subordinate leaders to spread innovative solutions within organizations and the Army at large. Organizational leaders actively coach and mentor subordinate leaders for their future leadership roles.

COMMUNICATES

Too often we place the burden of comprehension on those above or below us— assuming both the existence of a common language and a motivation.

General Edward C. Meyer
Chief of Staff, Army (1979-1983)

Ensuring Shared Understanding

11-18. Organizational leaders know themselves, the mission, and the message. They owe it to their organization and their subordinates to share as much information as possible. An open, two-way exchange of information reinforces sharing team values and signals constructive input is appreciated.

11-19. Communicating openly and clearly with superiors is important but critically important for organizational leaders. Understanding the superior's intent, priorities, and thought processes makes it easier to anticipate future planning and resource priorities. Understanding the azimuth of the higher headquarters reduces the amount of course corrections at the lower levels, thus minimizing friction and maintaining a stable organizational climate.

Leveraging the Staff as a Communications Tool

11-20. Organizational leaders constantly need to understand what is happening within their organization, developing laterally and unfolding within the next two higher echelons. Networking between staffs gives organizational leaders a broader picture of the overall operational environment and its dimensions. Coordination allows leaders to constantly interact and share thoughts, ideas, and priorities through multiple channels, creating a more complete picture. With reliable information, staffs can productively assist in turning policies, concepts, plans, and programs into achievable results and quality products.

11-21. By interacting with the next-higher staff, organizational leaders better understand the superior's priorities and impending shifts. This helps set the conditions for their own requirements and changes. Constantly sensing, observing, talking, questioning, and actively listening helps organizational leaders better identify and solve potential problems or to avoid them. It allows them to anticipate decisions and put their outfit in the best possible position in time and space to appropriately respond and execute.

Using Persuasion to Build Teams and Consensus

11-22. Persuasion is an important method of communication for organizational leaders. Well-developed skills of persuasion and openness to working through controversy in a positive way help organizational leaders overcome resistance and build support. These characteristics are important in dealing with other organizational leaders, multinational partners, and in the socio-political arena. By reducing grounds for misunderstanding, persuasion reduces wasted time in overcoming unimportant issues. It also ensures involvement of others, opens communication with them, and places value on their opinions—all critical team-building actions. Openness to discussing one's position and a positive attitude toward a dissenting view often defuses tension and saves time. By demonstrating these traits, organizational leaders also provide an example that subordinates can use in self-development.

11-23. In some circumstances, persuasion may be inappropriate. In combat, all leaders must often make decisions quickly, requiring a more direct style when leading and deciding on courses of action.

DEVELOPING

11-24. Comparatively, organizational leaders take a long-term approach to developing the entire organization. They prepare their organizations for the next quarter, next year, or even five years from now. The responsibility to determine how our Army fights the next war lies with today's Army leaders, especially those at the organizational and strategic levels. Leaders at the organizational level rely more on indirect leadership methods, which can make developing, leading, and achieving more difficult.

Creates a Positive Environment

> *It is not enough to fight. It is the spirit which we bring to the fight that decides the issue. It is morale that wins the victory.*
>
> General of the Army George C. Marshall
> Chief of Staff, Army (1939–1945)

11-25. An organization's climate springs from its leader's attitudes, actions, and priorities. These are engrained through choices, policies, and programs. Once in an organizational leadership position, the leader determines the organizational climate by assessing the organization from the bottom up. Once this assessment is done, the leader can provide clear guidance and focus (purpose, direction, and motivation) to move the organization towards the desired end state.

11-26. A climate that promotes the Army Values and fosters the Warrior Ethos encourages learning, promotes creativity and performance, and establishes cohesion. The foundation for a positive environment is a healthy ethical climate, although that alone is insufficient. Characteristics of successful organizational climates include a clear, widely known purpose; well-trained and confident Soldiers; disciplined, cohesive teams; and trusted, competent leaders.

11-27. To create such a climate, organizational leaders recognize mistakes as opportunities to learn, create cohesive teams, and reward leaders of character and competence with increasing responsibilities. Organizational leaders value honest feedback and constantly use all available means to maintain a feel for the organization within the contemporary operational environment. Special staff members who may be good sources for quality feedback include equal opportunity advisors, chaplains, medical officers and legal advisors. The organizational leader's feedback methods include town hall meetings, surveys, and councils.

PREPARES SELF

11-28. Leadership begins at the top, and so does developing. Organizational leaders keep a focus on where the organization needs to go and what all leaders must be capable of accomplishing. As visible role models, they continually develop themselves and actively counsel their subordinate leaders in professional growth. At the organizational level, commanders ensure that systems and conditions are in place for objective feedback, counseling, and mentoring for all the organization's members.

11-29. Self-aware organizational leaders who know their organizations generally achieve high quality results. Confident and competent organizational leaders do not shy away from asking their closest subordinates to give them informal feedback. This includes feedback about their leadership behaviors in critical training situations. It is all part of an open assessment and feedback effort. When they are part of official after-action reviews (AARs), organizational leaders should also invite subordinates to comment on how the leaders could have made things better. That is important since errors by organizational leaders are spotted easily and can often affect those they lead. Consequently, admitting, analyzing, and learning from these errors add value to the training. For the Army's organizational leaders—just like for leaders at other levels—reflecting, learning, and applying corrective actions in peacetime is critical for effectiveness in crisis.

11-30. While basic leader competencies stay the same across levels, moving from direct leader positions to the organizational level requires a shift in approach. Professional military education and Civilian Education System are designed to facilitate the transition in the scope and breadth of responsibilities. Leaders need to become accustomed to rely on less direct means of direction, control, and monitoring.

11-31. Developing as a leader does not only mean acquiring more skill, it also requires letting go of certain things. The demands on leaders vary at different levels. What may occupy a great deal of a leader's time at a lower level (for example, face-to-face supervision of Soldiers) may involve less time at higher levels. Certain technical skills that are vital for a direct leader may be of little importance to a strategic leader who needs to spend most of the time on strategic, system-wide leadership issues. As a result, leaders emphasize some skills less as the focus of leadership changes.

DEVELOPS OTHERS

11-32. One important organizational leader responsibility is to create an environment that enables and supports people within the organization to learn from their experiences and those of others. Operational leaders know they bear major responsibility for training the leadership of tomorrow's Army. They rely on an environment that leverages three sources of learning available throughout a Soldier and Army civilian's career: institutional training, education and training in operational assignments, and self-development through various procedures such as multisource assessment and feedback. To strengthen learning in organizations, organizational leaders can make four interdependent avenues available for lifelong learning: assignment oriented training, simulations, learning centers, and virtual training.

11-33. Effective organizational leaders develop leaders at all levels within their organization. Organizational leaders determine the potential of others. This takes special awareness of others and flexibility to build on strengths and address weaknesses. Developing others at this level is challenging. The organizational leader has to balance the criticality of the job, and who would do the best job, with the developmental needs of all subordinates.

11-34. Another consideration that organizational leaders take into account is how individuals learn and what they need to learn. Learning by trial and error and by making mistakes may be okay for some leaders, but others need to experience more successes than failures to develop self-confidence and initiative.

Building Team Skills and Processes

When a team outgrows individual performance and learns team confidence, excellence becomes a reality.

Joe Paterno
Head Coach, Penn State football

11-35. Organizational leaders recognize that the Army is a team, as well as a team of teams. As such, it is comprised of numerous functional organizations. These organizations are designed to perform necessary tasks and missions that in unison produce the effort of all Army components. At the mid-range, strategic leaders influence organizational leaders. As leaders of leaders, organizational leaders, in turn, influence subordinate leaders to achieve organizational goals.

11-36. Generally, organizational leaders rely on others to follow and execute their intent. Turning a battlefield vision or training goal into reality takes the combined efforts of many teams inside and outside of the organization. Organizational leaders build solid, effective teams by developing and training them.

11-37. Subordinates work hard and fight tenaciously when they are well trained and sense that they are part of a first-rate team. Collective confidence comes from succeeding under challenging and stressful conditions, beginning in training. Sense of belonging derives from experiencing technical and tactical proficiency—first as individuals and later collectively. That proficiency expresses itself in the confidence team members have in their peers and their leaders. Ultimately, cohesive teams combine into a network—a team of teams. Effective organizations work in synchronized fashion with teams to complete tasks and missions.

Encouraging Initiative and Acceptance of Responsibility

Never tell people how to do things. Tell them what to do and they will surprise you with their ingenuity.

General George S. Patton, Jr.
War As I Knew It (1947)

11-38. Since missions for larger organizations are more complex and involve many parallel efforts, leaders at higher levels must encourage subordinate initiative. Effective organizational leaders must delegate authority and support their subordinates' decisions while holding them accountable for their actions.

11-39. Successful delegation of authority involves convincing subordinates that they are empowered and have the freedom to act independently. Empowered subordinates understand that they bear more than the responsibility to get the job done. They have the authority to operate as they see fit, within the limits of commander's intent, assigned missions, task organization, and available resources. This helps them lead their people with determination.

11-40. Since delegation is a critical factor for success at the organizational level of leadership, leaders must know the character of their subordinates. Ultimate success may be in the hands of a properly empowered subordinate. Organizational leaders must know the resident talent within the organization and prepare subordinates to assume the critical roles when necessary. To empower the diverse elements within a larger organization, organizational leaders must also exploit the value of a creative staff composed of competent and trustworthy subordinates.

Choosing Talented Staff Leaders

11-41. A high-performing staff begins with putting the right people in the right positions. Organizational leaders are usually limited to find the right talents within their organization's resources. Nonetheless, they have choices on how to use good people. They thoughtfully select them from the entire organization—officers, noncommissioned officers and civilians who can think creatively. Organizational leaders make time to evaluate the staff and develop them to full capability with focused training. They avoid micromanaging the staff while trusting and empowering them to think creatively and provide truthful answers and feasible options.

11-42. One of the most important decisions for a commander is to select the right chief of staff or civilian deputy. By definition, the chief of staff or deputy is the principal assistant for directing, coordinating, supervising, and training the staff except in areas the commander reserves for himself. The chief of staff or deputy is a leader who has the respect of the team and can take charge of the staff, focus it, inspire it, and move it to achieve results in the absence of a commander. Although staff sections work as equals, it requires good chief of staff leadership to make them function as a cohesive team. (FM 6-0 discusses the role of the chief of staff.)

11-43. As leaders progress in their careers and the span of authority increases, incorporating subordinates in assessments and obtaining objective feedback from them becomes more important. Two proven techniques that involve subordinates in assessing are:

- In-process reviews.
- After-action reviews.

In-Process Review

11-44. An in-process review (IPR) is a quality control checkpoint on the path to mission accomplishment. Assessment begins with forming a picture of the organization's performance as early as possible. Leaders anticipate in which areas the organization might have trouble and focus attention there. Once the organization begins a mission, successive IPRs evaluate performance and give timely feedback. Leaders can use IPRs for major plans and operations as well as day-to-day events.

11-45. While IPRs are formal procedures, leaders should also consider informal methods of gathering feedback. Asking trusted subordinates for their candid input on leadership behaviors is another way leaders can assess their organization. Today's Soldiers are becoming tactically and technically knowledgeable at such a rapid pace that their feedback should not be discounted.

After-Action Review

11-46. AARs fulfill a similar role at the end of a mission. The AAR is a structured review process that allows participating Soldiers, leaders, and units to discover what happened during an event, why it happened, and how to do it better next time. Army leaders use AARs as opportunities to develop subordinates. When subordinates share in identifying reasons for success and failure, they become owners of how things are done. AARs also give invaluable opportunities to hear what is on subordinates' minds.

11-47. The key for meaningful AARs is for leaders to base reviews on accurate observations and exact recording of those observations. When evaluating a ten-day field exercise, good notes are essential to recall everything that happened. When recording observations, it is also helpful to look at things in a systematic way. Leaders can use a specific event or focus on an operating system to record observations. Most importantly, leaders must see things first-hand and not neglect tasks that call for subjective judgment such as unit cohesion, discipline, and morale. (FM 7-0 and FM 7-1 discuss training assessment.)

11-48. Inquisitive leaders who conduct regular assessments of themselves and their organizations hold their organizations to the highest standards. Open-minded reflection and corrective action in peacetime is critical for effective performance in crisis. Consider the 100-hour ground war of Operation Desert Storm. It was won before it was executed through hard work in countless field exercises, on ranges, and at the combat training centers. The continuous assessment process helped organizational leaders to translate critical peacetime lessons into decisive operations in war.

ACHIEVING

11-49. To get consistent results, organizational leaders have to be competent in planning, preparing, executing, and assessing. While leaders can continuously emphasize teamwork and cooperation, they also understand healthy competition can be an effective motivator. They must provide clear focus with their intent, so subordinates accomplish the mission, no matter what happens to the original plan.

Providing Direction, Guidance and Clear Priorities in Timely Manner

It is my duty to hear all; but, at last, I must, within my sphere, judge what to do, and what to forbear.

Abraham Lincoln
Letter to Charles D. Drake & Others
October 5, 1863

11-50. Organizational leaders are far more likely than direct leaders to be required to provide guidance and make decisions with incomplete information. Part of the organizational leaders' analysis must be to determine whether they have to decide at all, which decisions to make themselves, and which ones to push down to lower levels. While determining the right course of action, they consider possible second- and third-order effects and project far into the future—months or even years.

Accomplishing Missions Consistently

11-51. During operations, organizational leaders integrate and synchronize all available joint, interagency, and multinational resources. They assign specific tasks to accomplish the mission and empower their subordinates to execute within the given intent. The core strength for successfully executing the larger operational requirement centers on the leader's vision and the team's confidence and professionalism.

11-52. While a single leader in isolation can make good decisions, the organizational leader needs a creative staff to make quality decisions in an environment where operational momentum dominates a 24/7 cycle and decisions project into the future. In the complex operational environments faced today, organizational leaders must be able to rely on a creative and trustworthy staff to help acquire and filter huge amounts of information, monitor vital resources, synchronize systems, and assess operational progress and success.

11-53. Today's organizational leaders deal with a tremendous amount of information. Analysis and synthesis are essential to effective decision-making and program development. Analysis breaks a problem into its component parts. Synthesis assembles complex and disorganized data into a solution.

11-54. Good information management helps filter relevant information to enable organizational leaders and staffs to exercise effective command and control. Information management uses procedures and information systems to collect, process, store, display, and disseminate information. (FM 3-0 discusses information management. FM 6-0 discusses relevant information.)

11-55. Organizational leaders analyze systems and results to improve the organization and its processes. Performance indicators and standards for systems assist in analysis. Once organizational leaders complete an assessment and identify problems, they can develop appropriate solutions to address the problems.

MASTERING RESOURCES AND SYSTEMS

11-56. Organizational leaders must be masters of resourcing. Resources—including time, equipment, facilities, budgets, and people—are required to achieve organizational goals. Organizational leaders aggressively manage and prioritize the resources at their disposal to ensure optimal readiness of the organization. A leader's job is more difficult when unanticipated events such as emergency deployment shift priorities.

11-57. Organizational leaders are good stewards of their people's time and energy, as well as their own. They do not waste resources but skillfully evaluate objectives, anticipate resource requirements, and efficiently allocate what is available. They balance available resources with organizational requirements and distribute them in a way that best achieves organizational goals in combat or peacetime.

11-58. For example, in the early phases of an operation, airfields and supply routes to an area are often austere or nonexistent. Innovative logisticians coordinate available airlift, time-phasing cargo destined for forward operating bases. What takes priority? Bullets, food, water, fuel, personnel replacements, or mail? A good organizational leader will base prioritization decisions on multiple information sources: the warfighters' assessments, input from supporting units, personal situation assessments, and the commander's intent.

11-59. Because of the more indirect nature of their influence, organizational leaders continuously assess interrelated systems and design longer-term plans to accomplish missions. They must continuously sharpen their abilities to assess and balance their environments, organization, and people. Organizational leaders determine the cause and effect of shortcomings and translate these new understandings into workable plans and programs. They then allow subordinate leaders latitude to execute and get the job done.

11-60. Leaders who reach the organizational level must have developed a comprehensive systems perspective. This allows them to balance doctrine, organization, training, materiel, leadership and education, personnel, and facilities. Together with the Army Values and the Warrior Ethos, these systems provide the framework for influencing people and organizations at all levels. They are the foundation for conducting a wide variety of operations and continually improving the organization and the force.

Understanding and Synchronizing Systems

11-61. All leaders, especially organizational leaders, apply a systems perspective to shape and employ their organizations. The ability to understand and effectively leverage systems is critical to achieving organizational goals, objectives, and tasks. Organizational leadership, combined with effective information and systems management, can effectively generate combat power through superior logistics.

Leveraging the Logistics System to Increase Combat Power

During Operation Desert Shield (1990), a contingent of Army civilians deployed to a combat theater depot to provide critical warfighting supplies and operational equipment to the Third U.S. Army.

Two senior Army civilian leaders, the depot's deputy director of maintenance and the chief of the vehicle branch, confronted a critical issue: generating additional combat power. They had to devise a plan to replace the standard M1 tanks of several arriving units with upgraded M1A1s featuring greater firepower, better armor, and an advanced nuclear, biological, and chemical protective system. However, simple fielding was not enough. The civilian maintenance teams had to perform semiannual and annual maintenance checks, incorporate critical modifications such as applying additional armor, and repaint the tanks in desert camouflage patterns.

While some peacetime fieldings may take 18 to 24 months to complete, the two logistic leaders set an ambitious goal of 6 months. The team clearly understood the systems and resources that needed to be mobilized to get the job done to standard. Despite some initial skepticism, they completed the project in 2 months.

Success for the civilian logistic organization was solidly based on clear intent, firm objectives, systems knowledge, innovation, and leadership by example. Through a concerted effort, all critical combat units of Operation Desert Storm crossed the line of departure with confidence in the reliability and lethality of their weapons systems.

Synchronizing Tactical Systems

11-62. Organizational leaders must be masters of tactical and operational synchronization. They must arrange activities in time, space, and purpose to mass maximum relative combat power or organizational effort at a decisive point and time. Through synchronization, organizational leaders focus warfighting functions to mass the effects of combat power at the chosen place and time to overwhelm an enemy or to dominate a situation. The warfighting functions are—

- Intelligence.
- Movement and maneuver.
- Fire support.
- Protection.
- Sustainment.
- Command and control.

11-63. Organizational leaders at corps and above further synchronize by applying the complementary and reinforcing effects of all joint military and nonmilitary assets to overwhelm opponents at one or more decisive points. Effective synchronization requires leaders to pull together technical, interpersonal, and conceptual abilities and apply them to warfighting goals, objectives, and tasks.

11-64. The operational skill of synchronizing a series of tactical and operational events is demanding and far-reaching. The following example shows the complexity of an operation that synchronized joint, multinational, and civilian support assets for an evacuation operation for Americans and foreign nationals.

Joint and Multinational Synchronization during Operation Assured Response

For Operation Assured Response in Liberia, forces from the Republic of Georgia, Italy, and Germany joined with U.S. special operations, Air Force, Navy, and Marine forces to conduct a noncombatant evacuation operation. In early 1996, gunmen had filled the streets of Monrovia, Liberia as the country split into armed factions intent on seizing power. The situation worsened as faction members took hostages.

On 9 April 1996, President Clinton ordered the U.S. military to evacuate Americans and designated third party foreign nationals. In rapid response, the Army deployed special forces, an airborne company, signal augmentation, and a medical section as part of a special operations task force from Special Operations Command–Europe.

Army forces entered Monrovia's Mamba Point embassy district where they established security for international relief agencies headquartered there. Additional Army forces reinforced Marine guards at the American embassy and secured the central evacuee assembly collection point. Navy helicopters then flew the evacuees to Sierra Leone.

The combined capabilities of the Army, other Services, and multinational troops evacuated U.S. and foreign citizens from 73 countries from Liberia demonstrating the effectiveness and importance of synchronized joint, multinational operations.

Assessing to Ensure Mission Success and Organizational Improvement

11-65. Assessing situations reliably—and looking at the state of the organizations and their component elements—is critical for organizational leaders to achieve consistent results and mission success. Accurate assessment requires their instincts and intuitions based on the reliability of information and their sources. Quality organizational assessment can determine weaknesses and force focused improvements.

11-66. In addition to designing effective assessment systems, organizational leaders set achievable and measurable assessment standards. Assisted by their staffs, the chain of command, and other trusted advisors, leaders ensure that these are met. To get it right, organizational leaders ask:

- What is the standard?
- Does the standard make sense to all concerned?
- Did we meet it?
- What system measures it?

- Who is responsible for the system?
- How do we reinforce or correct our findings?

11-67. Because their decisions can have wide-ranging effects, leaders must be more sensitive how their actions affect the organization's climate. The ability to discern and predict second- and third-order effects helps organizational leaders assess the health of the organizational climate and provide constructive feedback to subordinates.

11-68. Attempting to predict second- and third-order effects may result in identifying resource requirements and changes to organizations and procedures. For instance, when the Army Chief of Staff approves a new military occupational specialty code for the Army, the consequences are wide-ranging. Second-order effects may mean specialized schooling, a revised promotion system for different career patterns, and requirements for more doctrinal and training material to support new specialties. Third-order effects include resource needs for training material and additional instructor positions at the appropriate training centers and schools. All leaders are responsible for anticipating the consequences of any action. Thorough planning and staff analysis can help, but anticipation also requires imagination, vision, and an appreciation of other people, talents, and organizations.

Chapter 12

Strategic Leadership

Final decisions are made not at the front by those who are there, but many miles away by those who can but guess at the possibilities and potentialities.

General Douglas MacArthur
Reminiscences (1964)

12-1. This chapter covers strategic leadership and puts the role of the strategic leader in perspective for all those who support strategic leaders. To support strategic leaders effectively—general and some senior field grade officers as well as senior Army civilians—one must fully understand the distinct environment in which these leaders work and the special considerations it requires.

12-2. Strategic leaders are the Army's ultimate multiskilled pentathletes. They represent a finely balanced combination of high-level thinkers, accomplished warfighters, and geopolitical military experts. Strategic leaders simultaneously sustain the Army's culture, envision the future force, and convey that vision to a wide audience. They often personally spearhead institutional change. Their leadership scope is enormous, typically responsible for influencing several thousand to hundreds of thousands of Soldiers and civilians.

12-3. To maintain focus, strategic leaders survey the environment outside the Army today to understand the context for the institution's future roles better. They use their knowledge of the current force to anchor their vision of the future force in reality-grounded analysis. Strategic leaders apply additional knowledge, experience, techniques, and skills beyond those required by direct and organizational leaders. In a strategic environment of extreme complexity, ambiguity, and volatility, strategic leaders must think in multiple time periods and apply more adaptability and agility to manage change. In addition to accepting harsher consequences for their actions, strategic leaders extend influence in an environment where they interact with many high level leaders over whom they have minimal formal authority or no authority at all.

12-4. Strategic leaders are experts in their own fields of warfighting and leading large organizations, and have to be comfortable in the departmental and political environments of the Nation's decision making. They have to deal competently with the public sector, the executive branch, and the legislature. America's complex national security environment requires an in-depth knowledge of the diplomatic, informational, military, and economic instruments of national power, as well as the interrelationships among them. Leaders not only know themselves and their own organizations, but also understand a host of different players, rules, and conditions.

12-5. Strategic leaders are keenly aware of the complexities of the national and international security environment. Their decisions take into account factors such as congressional hearings, Army budget constraints, Reserve Component issues, new systems acquisition, Army civilian programs, research, development, contracting, and inter-service cooperation. Strategic leaders process information from these areas quickly while assessing alternatives. Then they formulate practical decisions and garner support. Highly developed interpersonal abilities are essential to building consensus between civilian and military policy makers on national and international levels.

12-6. Strategic leaders need to understand organizational, national, and world politics. They operate in intricate networks of overlapping and sometimes competing constituencies. They participate in and shape endeavors extending beyond their span of responsibility. As institutional leaders, they represent their organizations to Soldiers, Army civilians, citizens, public officials, and the media, as well as to other Services and nations. Communicating effectively with different audiences is vital to any institution's success.

12-7. While direct and organizational leaders have a more near- and mid-term focus, strategic leaders must concentrate on the future. They spend much of their time looking toward long-term goals and positioning for long-term success even as they often contend with mid-term and immediate issues and crises. With that perspective and limited stabilization in their duty positions, strategic leaders seldom see their ideas come to fruition during their tenure.

12-8. To create powerful organizations and institutions capable of adapting and self-renew, strategic leaders and their staffs develop networks of knowledgeable individuals in organizations and agencies that can positively influence their own organizations. Through continuous assessments, strategic leaders seek to understand the personal strengths and weaknesses of the main players on a particular issue. Strategic leaders adeptly read other people while disciplining their own actions and reactions. Armed with improved knowledge of others, self-control, and established networks, strategic leaders influence external events by providing quality leadership, timely and relevant information, and access to the right people and agencies.

LEADING

Leadership is understanding people and involving them to help you do a job. That takes all of the good characteristics, like integrity, dedication of purpose, selflessness, knowledge, skill, implacability, as well as determination not to accept failure.

Admiral Arleigh A. Burke
Naval Leadership: Voices of Experience (1987)

12-9. When leading at the highest levels of the Army, the DOD, and the national security establishment, military, and Army civilian strategic leaders face highly complex demands from inside and outside the Army. The constantly changing world challenges their decision-making abilities. Despite the challenges, strategic leaders personally tell the Army story, make long-range decisions, and shape the Army culture to influence the force and its strategic partners within and outside the United States. They plan for contingencies across spectrum of conflicts and allocate resources to prepare for them, while constantly assessing emerging threats and the force's readiness. Steadily improving the Army, strategic leaders develop their successors, spearhead force changes, and optimize systems and operations while minimizing risk.

LEADS OTHERS

12-10. Strategic leaders act to influence both their organization and the external environment. Like direct and organizational leaders, strategic leaders lead by example and exert indirect leadership by communicating, inspiring, and motivating.

12-11. As noted earlier, strategic leaders develop the wisdom and reference framework to identify the information relevant to the situation. They also use their interpersonal abilities to develop a network of knowledgeable people in those organizations that can influence their own. They encourage staff members to develop similar networks. Through these formal and informal networks, strategic leaders actively seek information relevant to their organizations as well as subject matter experts who can assist them and their staffs. Using their networks, strategic leaders can call on the Nation's best minds and information sources because they may face situations where nothing less will suffice.

Providing Vision, Motivation, and Inspiration

A tremendous amount of work has been done to prepare the Army for the next century, but the job is not finished—and never will be. Change is a journey, not a destination….

General Gordon R. Sullivan Chief of Staff, Army (1991-1995)
Speech to the International Strategic Management Conference (1995)

12-12. The ability to provide clear vision is vital to the strategic leader, but forming a vision is pointless until the leader shares it with a broad audience, gains widespread support, and uses it as a compass to guide the organization. For the vision to provide purpose, direction, and motivation, the strategic leader must personally commit to it, gain commitment from the organization as a whole, and persistently pursue the goals and objectives that will spread the vision throughout the organization and make it a reality.

12-13. At the strategic level, leaders must ensure that their vision is clear to avoid confusion across joint and multinational forces. This allows each of them to turn an operational concept into their plans of action. On 14 November 1990, General Norman Schwarzkopf, Commander of U.S. Central Command, called 22 of his top commanders to Dhahran, Saudi Arabia to provide his vision and concept for Operation Desert Storm. The result was an example of clarity and simplicity.

From Vision to Victory

GEN H. Norman Schwarzkopf knew that this 14 November 1990 briefing was probably his most important during the planning phase for Desert Storm. He wanted to ensure that no one would leave with questions about the mission ahead.

He laid out his analysis of Iraq's forces: their force strength, their willingness to use chemical weapons, along with their weaknesses. He emphasized the strengths of his own forces and then revealed his vision. He laid out several objectives including destroying the Iraqi's capability as an effective fighting force. His message was clear—"destroy the Republican Guard."

One of Schwarzkopf's subordinate commanders reported in a later interview that it was "a mission that even privates could understand and one upon which they could all concentrate their efforts."

What had begun as a close-hold planning process was communicated horizontally and vertically so that each commander from division level and up heard the concept of operations from Schwarzkopf himself.

Schwarzkopf was pleased the President and Secretary of Defense gave him full authority to carry out his mission. In return, he stayed out of his commanders' way, allowing them to focus on their jobs and not be distracted by higher headquarters.

In mid-January 1991 when President Bush gave word to begin the operation, those tasked with carrying out the orders knew what their commander expected. The mission succeeded in driving the Iraqi occupying forces out, liberating Kuwait. Air superiority was gained and maintained, and much of Saddam Hussein's infrastructure and command and control were defeated during the conflict. Stability in the Gulf Region was regained, and the Republican Guard never fully recovered its fighting capability.

12-14. Strategic leaders identify trends, opportunities, and threats that could affect the Army's future and move vigorously to mobilize the talent that will help create strategic vision. In 1991, Army Chief of Staff General Gordon R. Sullivan formed a study group of two dozen people to help construct his vision for the future Army. In this process, General Sullivan considered authorship less important than shared vision:

> *Once a vision has been articulated and the process of buy-in has begun, the vision must be continually interpreted. In some cases, the vision may be immediately understandable at every level. In other cases, it must be translated—put into more appropriate language—for each part of the organization. In still other cases, it may be possible to find symbols that come to represent the vision.*

12-15. Strategic leaders are open to ideas from many sources, not just their own organizations. One such vision with long-term strategic and broad societal consequences was the integration of women into the armed forces.

Combat Power from a Good Idea

In 1941, as the American military was preparing for war, Congresswoman Edith Nourse Rogers correctly anticipated a manpower shortage in industry and in the armed forces as the military expanded. To meet growing needs, she proposed creating a Women's Army Auxiliary Corps (WAAC) of 25,000 women to fill administrative jobs, freeing men for service with combat units.

> After the United States entered the war, it became clear that the effort was on target but needed further expansion. Consequently, Congresswoman Rogers introduced another bill for 150,000 additional WAAC women. Although the bill met stiff opposition in some congressional quarters, a version of it passed. Eventually the Women's Army Corps was born and accepted as a major force multiplier.
>
> Congresswoman Rogers' vision of how to best satisfy the need for additional military personnel for a global war effort, significantly contributed to winning WWII and opened the door for employing the tremendous capabilities of female Soldiers.

12-16. The Army's institutional values-based culture affirms the importance of individuals and organizational quality through high standards, leader development, and lifelong learning initiatives. By committing to broad-based leader development, the Army often redefines what it means to be a Soldier. Army strategic leaders have consistently implemented changes to improve Soldier appearance together with performance standards. Introducing height and weight standards, raising physical fitness standards, embracing technology, and deglamorizing alcohol and tobacco have all contributed to fundamental changes in the Army's institutional culture.

EXTENDS INFLUENCE

12-17. Whether by nuance or overt presentation, strategic leaders vigorously and constantly represent the Army and its people by talking about what it is doing, and where it is going. Their audience is the Army itself, the Nation, and the rest of the world. There is a powerful responsibility to explain things to the American people, who support their Army with the essential resources of money and people. Whether working with federal agencies, the media, other countries' militaries, other Services, or their own organizations, strategic leaders rely increasingly on writing and public speaking to reinforce the Army's central messages.

12-18. Throughout the United States' history, strategic leaders have determined and reinforced the message that speaks to the soul of this Nation and unifies the armed forces. In 1973, Army leaders at all levels embraced "The Big Five" as a primary transformation message. It focused on the weapons systems that would help change the conscript Army into a professional volunteer force capable of dominating the Soviet threat. The Big Five included a new tank, an Infantry fighting vehicle, an advanced attack helicopter, a new utility helicopter, and an air defense system. Those programs soon yielded the M1 Abrams tank, the M2/M3 Bradley IFV, the AH-64 Apache, the UH-60 Blackhawk helicopter, and the Patriot missile system.

12-19. These modernization initiatives were more than just newer and better hardware; they improved concepts and doctrine on how to fight and win against a massive Soviet-style force. Soldiers experienced these improvements during schooling and in the field. The synergism of new equipment, new ideas, and good leadership ultimately resulted in the Army of Excellence.

12-20. Strategic leaders use focused messages to extend influence and to gain public support during crisis and war. An example of extending influence beyond the Army's sphere was Operation Desert Shield. During the deployment phase, strategic leaders decided to invite local reporters to the theater of war to focus reporting on mobilized Reserve Component units from local communities. The reporting focus had several positive effects. It conveyed the Army's deployment story to the citizens of hometown America, which resulted in a flood of mail from countless citizens to their deployed Soldiers. The most significant effect was soon felt by all Soldiers—a renewed pride in themselves and the Army.

12-21. Using Gulf War experiences, the Army's strategic leaders were able to improve their sharing of the Army's story during Operation Enduring Freedom and Operation Iraqi Freedom. Embedded reporters better informed the public about military culture while also giving the American people and the world a picture of the accomplishments of the Army during all phases of operations.

12-22. Often, strategic leadership beyond the traditional chain of command occurs when sending a symbolic message. Joshua Chamberlain's greatest contribution to our Nation may not have been at Gettysburg but at Appomattox. Then a major general, Chamberlain was chosen to command the parade at which General Lee's Army of Northern Virginia laid down their arms and colors. General Grant had directed a simple ceremony to recognize the Union victory without humiliating the Confederates. Chamberlain sensed the need for something even greater. Instead of gloating as the vanquished army passed, he directed his bugler to sound the commands for attention and present arms. His units came to attention and rendered a salute in respect. That act of military honor symbolized the beginning of a new era of respect, reconciliation, and reconstruction. It also highlighted a brilliant but humble leader, brave in battle and respectful in peace.

Negotiating Within and Beyond National Boundaries

12-23. Strategic leaders must often rely on negotiating skills to obtain the cooperation and support necessary to accomplish a mission or meet the command's needs. The North Atlantic Treaty Organization (NATO) provides many good examples. When NATO sent national contingents as part of the implementation force (IFOR) to Bosnia in response to the Dayton Peace Accords of 1995, all contingents had specific national operational limitations imposed on them. All contingent commanders maintained direct lines to their national governments to clarify situations immediately that may have exceeded those limits. Based on these political and cultural boundaries, NATO strategic leaders had to negotiate plans and actions that ordinarily would have required issuing simple orders. In the spirit of cooperation, commanders had to interpret all requirements to the satisfaction of one or more foreign governments.

12-24. The IFOR experience taught that successful negotiating requires a wide range of interpersonal skills. To resolve conflicting views, strategic leaders visualize several possible end states, while maintaining a clear idea of the best end state from the national command's perspective. Sometimes strategic leaders must also use tact to justify standing firm on nonnegotiable points while still communicating respect for other participants.

12-25. A successful negotiator must be particularly skilled in active listening. Other essential personal characteristics include good judgment and mental agility. Negotiators must be able to diagnose unspoken agendas and detach themselves from the negotiation process. Successful negotiating also involves communicating a clear position on all issues while conveying a willingness to bargain on negotiable portions. This requires recognizing what is acceptable to all concerned parties and working towards a common goal.

12-26. To reach acceptable consensus, strategic leaders often circulate proposals early so that further negotiations can focus on critical issues and solutions. Confident in their abilities, strategic leaders do not claim every good idea. Their commitment to selfless service enables them to subordinate the need for personal recognition to finding positive solutions that produce the greatest good for their organization, the Army, and the Nation.

Building Strategic Consensus

12-27. Strategic leaders are skilled at reaching consensus and building coalitions. They may apply these skills to tasks—such as designing combatant commands, joint task forces, and policy working-groups—or determine the direction of a major command or the Army as an institution. Strategic leaders routinely bring designated people together for missions lasting from a few months to years. Using peer leadership rather than strict positional authority, strategic leaders carefully monitor progress toward a visualized end state. They focus on the health of the relationships necessary to achieve it. Interpersonal contact sets the tone for professional relations: strategic leaders must be tactful.

12-28. General Dwight D. Eisenhower's creation of Supreme Headquarters Allied Expeditionary Force (SHAEF) during World War II is an inspiring example of coalition building and sustaining fragile relationships. General Eisenhower exercised his authority through an integrated command and staff structure that respected the contributions of all nations involved. To underscore the united team spirit, sections within SHAEF had chiefs of one nationality and deputies of another.

12-29. Across the Atlantic Ocean, General George C. Marshall, the Army Chief of Staff, also had to seek strategic consensus with demanding peers, such as Admiral Ernest J. King, Commander in Chief, U.S. Fleet and Chief of Naval Operations. General Marshall expended great personal energy ensuring that inter-Service feuding at the top did not dilute the U.S. war effort. Admiral King, a forceful leader with strong and often differing views, responded in kind. Because of their ability to find consensus, President Franklin D. Roosevelt had few issues of major consequence to resolve once he had issued a decision and guidance.

Leads By Example

Nothing valuable can be lost by taking time. If there be an object to hurry any of you in hot haste to a step which you would never take deliberately, that object will be frustrated by taking time; but no good object can be frustrated by it.

Abraham Lincoln
First Inaugural Address
March 4, 1861

12-30. Strategic leaders have great conceptual resources, including an intellectual network to share thoughts and plan for the institution to have continued success and well-being. Decisions made by strategic leaders—whether combatant commanders deploying forces or service chiefs initiating budget programs—often result in a major commitment of resources. Once in motion these are expensive and tough to reverse. Therefore, strategic leaders rely on timely feedback throughout the decision-making process to avoid making a final decision based on inadequate or faulty information. Their purpose, direction, and motivation flow down while information and recommendations surface from below. Strategic leaders leverage the processes of the DOD, Joint Staff, and Army strategic planning systems to provide purpose and direction to subordinate leaders. These systems include—

- The Joint Strategic Planning System.
- The Joint Operation Planning and Execution System.
- The Planning, Programming, and Budgeting System.

12-31. No matter how many systems are involved or how complex they are, providing motivation for mission accomplishment remains a primary responsibility of the strategic leader. Since strategic leaders are constantly involved in planning and because decisions at their level are often complex and depend on numerous variables, there can be a temptation to overanalyze. Their conscientiousness, knowledge, competence, judgment, and agility help them know when to decide. A strategic leader's decision at a critical moment in wartime can rapidly alter the course of an entire campaign.

Leading and Inspiring Institutional Change

If you don't like change, you're going to like irrelevance even less.

General Eric Shinseki
Chief of Staff, Army (1999-2003)

12-32. To fulfill its mission, the Army must be able to deal with inevitable change. The Army's strategic leadership recognizes that as an institution, the Army is in a nearly constant state of flux: processing and integrating new people, new missions, new technologies, new equipment, and new information. The challenge for the strategic leaders is to create grounded future-oriented change.

12-33. Strategic leaders lead change by—

- Identifying the force capabilities necessary to accomplish the National Military Strategy.
- Assigning strategic and operational missions, including priorities for allocating resources.
- Preparing plans for using military forces across the full spectrum of operations.
- Creating, resourcing, and sustaining organizational systems, including:
  - Conducting force modernization programs.
  - Requisite personnel and equipment resources.
  - Essential command, control, communications, computers, and intelligence systems.
- Developing and improving doctrine as well as the training methods supporting it.
- Planning for the second- and third-order effects of change.
- Maintaining an effective leader development program and other human resource initiatives.

12-34. Strategic leaders accept change in proactive, not in reactive fashion. They anticipate change even as they shield their organizations from unimportant and bothersome influences. The history of the post-Vietnam volunteer Army illustrates how strategic leaders can effectively shape change to improve the institution while continuing to deal with routine operations and requirements.

Change after Vietnam

The Army began seeking only volunteers in the early 1970s. Transforming to an all-volunteer force required a multitude of doctrinal, personnel, and training initiatives that took years to mature. While transforming, the Army tackled societal changes, such as drug abuse, racial tensions, and a sagging economy. Simultaneous with changes in the personnel arena, new equipment, weapons, vehicles, and uniforms further enhanced capabilities and readiness.

With vision, confidence, and personal example, the Army's strategic leaders succeeded in overhauling Army doctrine to create an environment that improved training at all levels. New training management doctrine and the Combat Training Center programs provided a solid foundation for uniformly understood warfighting.

All these changes required ambitious, long-range plans and aggressive leader actions. The result was the Army of Desert Storm, a force greatly different from the force of fifteen years earlier. The Army's transformation did not happen by accident. The blueprint for change came from the minds of strategic leaders who had taken the lessons of the past and combined them with a vision for the future. It was then put into reality and ultimately battlefield victory—by the hard work of direct and organizational leaders, as well as all member of the Army team.

12-35. Generally, strategic leaders know that institutional change requires influence grounded in commitment rather than forced compliance. Commitment must be reinforced consistently throughout the multiple levels of the organization. While all levels of leaders lead change, strategic level leaders make the most-sweeping changes and ones that focus on the most distant horizon. Strategic leaders guide their organizations through eight distinct steps if their initiatives for change are to make lasting progress. The critical steps of the leading change process are:

- Demonstrate a sense of urgency by showing both the benefits and necessity for change.
- Form guiding coalitions to work the process of change from concept through implementation.
- With the guiding coalitions and groups, develop a vision of the future and strategy for making it a reality.
- Clearly communicate the future vision throughout the institution or organization; change is most effective when all members embrace it.
- Empower subordinates at all levels to pursue widespread, parallel efforts.
- Plan for short-term successes to validate key programs and keep the vision credible.
- Consolidate the successful programs to produce further change.
- Ensure that the change is culturally preserved.

12-36. The result is an institution that constantly prepares for and shapes the future environment. Strategic leaders seek to sustain the Army as that kind of institution.

Displaying Confidence in Adverse Conditions—Dealing with Uncertainty and Ambiguity

Difficulties mastered are opportunities won.

Sir Winston Churchill
Prime Minister of Great Britain, WW II

12-37. Strategic leaders operate in an environment of increased volatility, complexity, and ambiguity. Since change may arrive unexpectedly, strategic leaders prepare intellectually for a range of threats and scenarios. Planning and foresight cannot predict or influence all future events, strategic leaders work carefully to shape the future with the means available to them through the diplomatic, informational, military, and economic instruments of national power, as well as their character, competence, and confidence.

12-38. Strategic leaders best deal with complexity by embracing it. This means they expand their frame of reference to fit a situation rather than reducing a situation to fit their preconceptions. Because of their sense of duty, competence, intellectual capacity, and wise judgment, they tolerate ambiguity, as they will never have all the information they want. Instead, strategic leaders carefully analyze events and decide when to make a decision, realizing that they must innovate and accept some risk. Once they make decisions, strategic leaders explain them to their organization, the Army, and the Nation.

12-39. In addition to demonstrating the agility required to handle competing demands, strategic leaders understand complex cause-and-effect relationships and anticipate the second- and third-order effects of their decisions throughout the organization. Some second- and third-order effects are desirable and leaders can purposely pursue specific actions to achieve them. While the highly volatile nature of the strategic environment may tempt some strategic leaders to concentrate mainly on the short term, they cannot allow the crisis of the moment to absorb them. They must remain focused on their responsibility to shape an organization or policies that will perform successfully over the next ten to twenty years.

COMMUNICATES

12-40. Communication at the strategic level is very encompassing. It involves a wide array of staffs and many functional and operational components interacting with each other, as well as with external agencies. These complex, information-sharing relationships require strategic leaders to employ comprehensive communications skills when representing their organizations. One prominent difference between strategic leaders and leaders at other levels is the greater emphasis on symbolic communication. The example that strategic leaders set—their words, decisions, and their actions—have meaning beyond their immediate consequences, more than those of direct and organizational leaders.

12-41. Strategic leaders must identify those actions that transmit messages and carefully use their positions of prominence and authority to convey them to the right target audiences. Strategic leaders generally send messages of a broader scope that support traditions, the Army Values, or a particular program. The broad scope also helps strategic leaders indicate their priorities and direction. To influence those audiences, strategic leaders must simultaneously convey professional integrity and confidence in the message to earn general trust. As General George C. Marshall noted, strategic leaders become experts in the art of persuasion.

12-42. To achieve the desired effect, strategic leaders commit to a few common, powerful, and consistent messages, which they repeat in different forms and settings. They devise and follow a communications campaign plan—written or conceptual—outlining how to deal with each target group. When preparing to address a specific audience, strategic leaders determine its composition and agenda beforehand so they know how best to reach its members. They carefully assess the impact of the message in the categories of medium, frequency, specific words, and the general environment. It is essential to ensure the message is going to all the right groups with the desired effect.

12-43. One form of communication strategic leaders must use effectively to persuade individuals, rather than groups, is dialogue. Dialogue is conversation that takes the forms of advocacy and inquiry. Advocacy seeks to advance a position, and inquiry looks to find out more about another's position or perspective. Dialogue that blends the two has greater value for leaders who must deal with issues, which are more complex than personal experience. To advocate a view, leaders make reasoning explicit, invite others to consider the view, encourage others to provide different views, and explore how views differ. When inquiring into another's view, leaders should voice their assumptions and seek to identify what evidence or support exists for the other view. Open dialogue can help overcome reluctance to consider different points of view.

DEVELOPING

12-44. Strategic leaders make institutional investments with a long-term focus. Their fundamental goal is to leave the Army better than they found it. This effort calls for the courage to experiment and innovate. Developing the institution, its organizations, and people involves an ongoing tradeoff between operating today and building for tomorrow. Strategic leaders apply wisdom and a refined frame of reference to understand what traditions should remain stable and which long-standing methods need to evolve. Most importantly, strategic leaders set the conditions for long-term success of the organization by developing subordinates who can take the institution to its next level of capability.

CREATES A POSITIVE ENVIRONMENT TO POSITION THE INSTITUTION FOR THE FUTURE

A good soldier, whether he leads a platoon or an army, is expected to look backward as well as forward; but he must think only forward.

General Douglas MacArthur
Graduation Speech at the United States Military Academy (17 June 1933)

12-45. The Nation expects military professionals as individuals and the Army as an institution to learn from the experience of others, apply that learning to understanding the present, and prepare for the future. Such learning requires both individual and institutional commitments. Strategic leaders, by personal example and critical resourcing decisions, sustain the culture and policies that encourage both the individual and the Army to learn and evolve.

12-46. Like organizational and direct leaders, strategic leaders must model character with all their actions. Only experience can validate the Army Values. Subordinates will know of the Army Values after seeing those around them actually live by them.

12-47. Strategic leaders ensure the Army Values and the Warrior Ethos remain fundamental to the Army's institutional culture. The culture affects how they act in relation to each other and towards outside agencies, as well as how they approach the mission. A solid and values-based culture helps define the boundaries of acceptable behavior, ranging from how to wear the uniform to how to interact appropriately with foreign cultures. It helps determine how people approach problems, make judgments, determine right from wrong, and establish proper priorities. Culture shapes Army customs and traditions through doctrine, policies, and regulations, and the philosophy that guides the institution. Professional journals, historical works, ceremonies—even the folklore of the organization—all contain evidence of the Army's institutional culture.

12-48. A healthy culture is a powerful motivational tool. Strategic leaders leverage it to guide and inspire large and diverse organizations. They use the institutional culture to support vision, accomplish the mission, and improve the organization. A cohesive culture molds the organization's morale, reinforcing an ethical climate solidly resting on the Army Values.

12-49. Strategic leaders promote learning by emplacing systems for studying the force and future environments. They resource a structure that constantly reflects on how the Army fights and what success requires. It requires constantly assessing the culture and deliberately encouraging creativity and learning.

12-50. Strategic leaders focus research and development efforts on achieving joint, interagency, and multinational synergy for success. They also coordinate time lines and budgets so that compatible and mutually supporting systems are fielded together.

12-51. Strategic leaders are also concerned that evolving forces have optimal capability over time. They prepare plans to integrate new equipment and concepts into the force as soon as components are available rather than waiting for all elements of a system to be ready before fielding it. Rehearsing the integration of systems or their separate components is often done during especially designed exercises to gain early feedback. The Louisiana Maneuvers in 1941 taught the Army what mechanized warfare would look like and how to prepare for it. The success of U.S. mechanized warfare validated most of the lessons learned. A study bearing the same name 50 years later advanced the conceptual Force XXI recreating the 4th Infantry Division as the first digitized division. Strategic leaders commissioned these forward-looking projects because the Army is dedicated to learning about operations in new environments and against evolving threats. The Louisiana Maneuvers can be seen as strategic counterparts to the rehearsals conducted at lower levels by direct and organizational leaders prior to an upcoming mission.

12-52. Strategic leaders are at the forefront of making the Army a lifelong learning organization, embracing the entire Army—Regular Army and Reserve Components as well as Army civilians. Modern strategic leaders must use the constantly evolving information technology and distributed learning, thus turning many institutions into classrooms without walls. The overarching goal is to provide the right education and training and to incorporate the best ideas rapidly into doctrine that ultimately improve and refine operational readiness.

PREPARES SELF WITH STRATEGIC ORIENTATION

12-53. All self-aware Army leaders build a personal frame of reference from schooling, experience, self-study, and assessment while reflecting on current events, history, and geography. Strategic leaders create a comprehensive frame of reference that encompasses their entire organization and places it in the strategic environment. To construct a useful framework, strategic leaders are open to new experiences and to comments from others including subordinates. Strategic leaders are reflective, thoughtful, and unafraid to rethink past experiences in order to learn from them. They are comfortable with the abstractions and concepts common in the operational and strategic environments. They try to understand the circumstances surrounding them, their organization, and the Nation.

12-54. Much like intelligence analysts, strategic leaders look at events and see patterns to determine when to intervene or act. A strategic leader's broad frame of reference helps identify the information most relevant to a strategic situation and find the heart of a matter without distraction. Cognizant strategic leaders with comprehensive frames of reference, and the wisdom that comes from experience and mental agility, are equipped to deal with events with complex causes. They can envision creative and innovative solutions.

12-55. A well-developed frame of reference also gives strategic leaders a thorough knowledge of organizational subsystems and their interacting processes. Cognizant of the interactive relationships among systems, strategic leaders foresee the possible effects of one system as it could affect the actions in others. That vision helps them anticipate and prevent potential problems.

Expanding Knowledge in Cultural and Geopolitical Areas—Mastering Strategic Art

The crucial military difference (apart from levels of innate ability) between Washington and the commanders who opposed him was that they were sure they knew all the answers, while Washington tried every day and every hour to learn.

James Thomas Flexner
George Washington in the American Revolution (1968)

12-56. Strategic leaders create their work on a broad canvas that requires broad technical skills and mastery of strategic art. Broadly defined, strategic art is the skillful formulation, coordination, and application of ends, ways, and means to promote and defend the national interest. Masters of the strategic art competently integrate the three roles performed by the complete strategist:

- Strategic leader.
- Strategic practitioner.
- Strategic theorist.

12-57. Using their understanding of the systems within their own organizations, strategic leaders work through the complexity and uncertainty of the strategic environment and translate abstract concepts into concrete actions. Proficiency in the science of leadership theory, programs, schedules, and systems, helps organizational leaders succeed. For strategic leaders, the intangible qualities of leadership draw on their long and varied experience to produce a rare art.

12-58. By reconciling political and economic constraints with the Army's needs, strategic leaders navigate to move the force forward using a combination of strategy and budget processes. They spend a great deal of time obtaining and allocating resources and determining conceptual directions, especially those judged critical for future strategic positioning and others necessary to prevent readiness shortfalls. They also oversee the Army's responsibilities under Title 10 of the United States Code.

12-59. Strategic leaders focus not so much on internal processes, but as to how the organization fits into DOD and the international arena. They ask broad questions, such as—

- What are the relationships among external organizations?
- What are the broad political and social systems in which the organization and the Army must operate?

12-60. Because of the complex reporting and coordinating relationships, strategic leaders must fully understand their roles, the boundaries of these roles, and the expectations of other departments and agencies. Understanding those interdependencies outside the Army helps strategic leaders do the right thing for the programs, systems, and people within the Army as well as for the Nation.

12-61. A strategic and institutional challenge occurred in the summer of 1990. While the Army was in the midst of the most precisely planned force drawdown in its history, Army Chief of Staff Carl Vuono had to halt the process to meet a crisis in the Persian Gulf. On short notice, General Vuono was required to call up, mobilize, and deploy the forces necessary to meet the immediate Gulf crisis while retaining adequate capabilities in other theaters. After the successful completion of Operation Desert Shield and Operation Desert Storm, he then redeployed the Third U.S. Army in 1991, demobilized the activated reserves, and resumed downsizing the Army to the smallest active force since the 1930s. By doing so without a major degradation of readiness, General Carl Vuono demonstrated mastery of the technical component of the strategic art.

Self-Awareness and Recognition of Impact on Others—Drawing on Conceptual Abilities

From an intellectual standpoint, Princeton was a world-shaking experience. It fundamentally changed my approach to life. The basic thrust of the curriculum was to give students an appreciation of how complex and diverse various political systems and issues are....The bottom line was that answers had to be sought in terms of the shifting relationships of groups and individuals, that politics pervades all human activity, a truth not to be condemned but appreciated and put to use.

Admiral William J. Crowe
Chairman, Joint Chiefs of Staff (1985-1989)

12-62. Strategic leaders, more so than direct and organizational leaders, draw on their self-awareness and conceptual abilities to comprehend and manage their more complex environments. Their environmental challenges include national security, theater strategies, operating in the strategic and theater contexts, and helping vast, complex organizations evolve. The variety and scope of strategic leaders' concerns demand the application of more sophisticated concepts and wisdom beyond pure knowledge.

DEVELOPS LEADERS

Certainly one of the reasons for [General] Marshall's success as a leader was not only his personal determination to learn but also his desire to share the knowledge he gained with his associates and subordinates, regardless of rank. He did this eagerly and willingly, without thought of personal glory, for the benefit of a common cause.

Edgar F. Puryear, Jr.
Nineteen Stars: A Study in Leadership (1971)

12-63. Strategic leaders develop subordinates through coaching, through providing policies and resources, and sharing the benefit of their perspective and experience (mentoring). To bridge the knowledge gap between organizational and strategic leaders, experienced strategic leaders can help newcomers by introducing important players and pointing out the critical places and activities. Strategic leaders become enablers as they underwrite the learning, efforts, projects, and ideas of rising leaders. Through developing others, strategic leaders help build a team of leaders prepared to fill critical positions in the future.

Counseling, Coaching, and Mentoring

12-64. More than a matter of following formats and structured sessions, mentoring by strategic leaders means giving the right people an intellectual boost so that they make the leap to successfully operating and creatively thinking at the highest levels.

12-65. Since few formal leader development programs exist beyond the senior service colleges, strategic leaders pay special attention to their subordinates' self-development. Leaders coach them on what to study, where to focus attention, whom to study as examples, and how to proceed along their career path. To impart their wisdom beyond coaching and mentoring, leaders speak to audiences at service schools about what happens at their level and share their perspectives with those who have not yet reached the highest levels of Army leadership. Today's subordinates will become the next generation of strategic leaders.

Building Team Skills and Processes

12-66. Given a more rapid transfer speed for all types of information, today's strategic leaders often have less time to assess situations, make plans, prepare an appropriate response, and execute for success. A world strategic environment in constant flux has increased the importance of building courageous, honest, and competent staffs and command teams.

12-67. Strategic leaders mold staffs and organizational teams, able to package concise, unbiased information and build networks across organizational lines. It is because strategic leaders make so many wide-ranging and interrelated decisions that they must be able to rely on imaginative staff members and subordinate leaders who comprehend the environment, foresee consequences of various courses of action, and identify crucial information.

12-68. Because they must be able to compensate for their own weaknesses, strategic leaders cannot afford to have staffs that blindly agree with everything they say. Strategic leaders encourage staffs to participate in open dialogue with them, discuss alternative points of view, and explore all facts, assumptions, and implications. Such dialogue assists strategic leaders in fully assessing all aspects of an issue and helps clarify their vision, intent, and guidance. As strategic leaders build and use effective staffs, they continually seek honest and competent people: Soldiers and civilians of all diverse backgrounds.

Assessing Developmental Needs and Foster Job Development

12-69. What strategic leaders do for individuals they mentor, they also seek to provide for the force at large. By committing money to select programs and projects or investing additional time and resources to specific actions, strategic leaders can set priorities. Ultimately, the Soldiers and civilians who develop those ideas become trusted assets themselves. Strategic leaders can choose wisely the ideas that bridge the gap between today and tomorrow and skillfully determine how best to resource critical ideas and people.

12-70. Living with time and budget constraints, strategic leaders must make difficult decisions about how much institutional development suffices. They can calculate how much time it will take to develop and nourish the Army's leaders and ideas for the future. They balance today's operational requirements against tomorrow's force structure and leadership needs. Their goal is to develop a core of Army leaders with the relevant competencies to steer the force into the future.

12-71. Programs like Training with Industry, advanced civil schooling, and foreign area officer education complement the training and education available in the Army's schools and contribute to shaping the people who will shape the Army's future. Strategic leaders develop the institution by using available Army resources. They skillfully complement this effort with resources offered by other Services or the public sector.

12-72. After the Vietnam War, the Army's leadership acknowledged that investing in officer development was so critically important that new courses were developed to revitalize the professional education for the force. Establishing the Training and Doctrine Command revived Army doctrine as a central intellectual pillar of the service. The Goldwater-Nichols Act of 1986 provided similar attention and increased emphasis on professional joint education and doctrine.

12-73. Likewise, there has been a huge investment in developing professional Army noncommissioned officer (NCOs). In 1973, the Army established the U.S. Army Sergeants Major Academy. This school became the pinnacle of formal military schooling for Army NCOs.

12-74. Complementary to military education systems for officers, warrant officers and NCOs, the civilian education system is the Army's program for developing Army civilian leaders. It continues throughout an individual's career as a lifelong learning initiative. The civilian education system provides a progressive, sequential, and competency-based leader development educational program beginning at entry level and continuing through managerial level. Senior service college is the apex of a civilian's leader development education and prepares civilians for positions that require an understanding of complex policy and operational challenges and increased knowledge of the national security mission. The Defense Leadership and Management Program is a comprehensive program of education and development for senior DOD civilian leaders with a DOD-wide perspective; substantive knowledge of the national security mission; a shared understanding, trust, and sense of mission with military counterparts; and strong leadership and management skills. Together, these programs provide Army civilians the requisite educational development opportunities, paralleling that of their counterpart in uniform.

ACHIEVING

Continuity and change are important in the life and vitality of any organization.... We achieve a healthy balance [by] maintaining continuity and creating change.

General John A. Wickham, Jr.
Chief of Staff, Army (1983-1987)

12-75. The National Security Strategy and National Military Strategy guide strategic leaders as they develop visions for their organizations. Strategic leaders must define for their diverse organizations what success means when executing to pursue their vision. They monitor progress and results by drawing on personal observations, organized review and analysis, strategic management plans, and informal discussions with Soldiers and Army civilians.

PROVIDING DIRECTION, GUIDANCE, AND CLEAR VISION

12-76. When providing direction, giving guidance, and setting priorities, strategic leaders must judge realistically what the future may hold. They incorporate new ideas, new technologies, and new capabilities. From a mixture of ideas, facts, conjecture, and personal experience, they create an image of what their organizations need to be and where it must go to get desired results.

12-77. The strategic leader's vision provides the ultimate sense of purpose, direction, and motivation for everyone in the organization. It is the starting point for developing specific goals and plans, a yardstick for measuring organizational accomplishment, and a check on organizational values. A shared vision throughout the organization is important for attaining commitment to change. A strategic leader's vision for the organization may have a time horizon of years, or even decades.

12-78. Strategic leaders seek to keep their vision consistent with the external environment, alliance goals, the National Security Strategy, National Defense Strategy, and National Military Strategy. Subordinate leaders align their visions and intent with their strategic leader's vision. A strategic leader's vision is in everything from small actions to formal written policy statements.

12-79. Regularly published concept papers creatively array future technologies and force structure against emerging threats. While no one can see in minute detail what the future force will look like exactly, the papers provide a snapshot for future options.

STRATEGIC PLANNING AND EXECUTION

12-80. Strategic-level plans must balance competing demands across the vast structure of DOD. The fundamental requirements for strategic-level planning are the same as planning at the direct and organizational levels. At all levels, leaders establish realistic priorities and communicate decisions. What adds complexity at the strategic level is the sheer number of players and resource factors that can influence the organization.

12-81. In the following excerpt, Paul R. Howe, former U.S. Army Master Sergeant, describes key planning, practice, and training strategies important to the successful planning for and execution of a mission.

The Quickest and Most Efficient Way to Plan

Alert the Force: Ensure that the system you use to alert and gather the force works and is simple and effective. Have a primary and secondary method of communication available.

Report: All members should report in, and leaders should be given a brief mission statement that can be passed on to all team members.

Begin Planning: After briefing their teams, team leaders should report to a central planning area to receive an information update and command guidance.

Pass Information On and Prepare Rehearsal Areas: Information should be passed down to the end users (teams) as rapidly as possible. Certain teams may be responsible for getting together a rehearsal site.

Brief the Plan: Once the plan is formulated and the commander gives the go-ahead available assets should be assembled and briefed as a group, if at all possible, to look at common danger and problem areas that need to be addressed.

Rehearse: Rehearsals should be conducted from the bottom level up and then the actions on the objective first. Hopefully, once assistant team leaders get the word of the mission and get their gear ready, they should have taken their individual teams and started to do walkthrough rehearsals on their own. They can walk through and discuss exterior movement, breach point procedures, interior movement, close-quarter battle procedures, and consolidation procedures for the target. They can also review medical procedures and do an equipment shakedown to ensure all the mission's essential gear is present and functioning.

Fix Problems: Rehearsals should start with all teams working on how they are going to take down the target, or actions on the objective. Then the rehearsal staff should throw in some problems to make the force think and ensure their backup plans work and are viable. Once this is accomplished, you can then work on your movement phase and on putting the entire package together.

Re-Rehearse, Re-Brief, and Integrate All Assets into the Plan: If there are too many problems with the initial plan, leadership should re-brief the entire plan to the group. Too many changes will confuse everyone, including key leaders, as to what the current plan is. Taking the time to re-brief will save everyone a world of headaches.

Prepare for Mission Execution: Once final rehearsals are complete, the force can catch their breath and stand by for the execution phase. It may come soon, or the action may take days. Leaders should act accordingly and ensure that the men are not burned out by leaning on the edge too long and too hard. Take a down time as appropriate. Then re-rehearse actions as needed to keep the force sharp.

12-82. The shift from Cold War to regional conflicts and the War on Terrorism within a decade demonstrates that the character of war is continuously changing. Strategic leaders must therefore constantly seek current information about the shifting strategic environment to determine what sort of force to prepare.

12-83. Questions strategic leaders must consider are:

- Where is the next threat?
- Will we have allies or contend alone?
- What will our national and military goals be?
- What will the exit strategy be?

12-84. Strategic leaders must be able to address the technological, leadership, and ethical considerations associated with conducting missions on an asymmetrical battlefield and typified by operations in Iraq and Afghanistan after the collapse of the original power structure. Strategic leaders will find themselves more than ever at the center of the tension between traditional warfare and the newer kinds of multiparty conflict emerging outside the industrialized world.

Allocating the Right Resources

12-85. Because lives are precious and materiel is scarce, strategic leaders must make tough decisions about priorities. Strategic Army priorities focus on projecting Landpower: the ability—by threat, force, or occupation—to promptly gain, sustain, and exploit control over land, resources, and people (see FM 1).

12-86. When planning for tomorrow, strategic leaders consistently call on their understanding and knowledge of the budgetary process to determine which combat, combat support, and combat service support technologies will provide the leap-ahead capability commensurate with the cost. Visionary Army leaders of the 1970s and 1980s realized that superior night-fighting systems and greater standoff ranges would expose fewer Soldiers to danger, yet kill more of the enemy. Those leaders committed the necessary money to developing and procuring these and other superior systems and ancillary equipment. Operation Desert Storm ultimately validated their vision and decisions when U.S. tanks destroyed Soviet-style armor before it could close within its own maximum effective range.

Leveraging Joint, Interagency, and Multinational Assets

12-87. Strategic leaders oversee the relationship between their organizations, as part of the Nation's total defense force, and the national policy apparatus. Among their continuous duties, strategic leaders—

- Provide military counsel in national policy forums.
- Interpret national policy guidelines and directions.
- Plan for and maintain the military capability required to implement national policy.
- Present the organization's resource requirements.
- Develop strategies to support national objectives.
- Bridge the gap between political decisions made as part of national strategy and the individuals and organizations that must carry out those decisions.

12-88. Just as direct and organizational leaders consider their sister units and support agencies, strategic leaders consider and work with other Services and government agencies. Consider that most of the Army's four-star billets today are joint or multinational. Almost half of the lieutenant generals hold similar positions on the Joint Staff, with the DOD, or in combatant commands. While the remaining strategic leaders are assigned to nominally single service organizations (Forces Command, Training and Doctrine Command, Army Materiel Command), they frequently work outside Army channels. In addition, many Army civilian strategic leaders hold positions that require a well-rounded joint perspective.

12-89. The complexity of joint and multinational requirements is two-fold. First, communication is more complicated because of different interests, cultures, and languages of multinational participants. Even the cultures and jargon of the other Services can differ. Second, subordinates may not be subordinate in the same sense as they are in a purely Army organization. Strategic leaders and their forces may fall under international operational control but retain their allegiances and lines of authority to their own national commanders. United Nations and NATO commands, such as IFOR, as well as cooperative arrangements between NATO or multinational forces in Operations Enduring Freedom and Iraqi Freedom, are examples of these complex arrangements.

Operating and Succeeding in a Multicultural Context

12-90. Creating a third culture is often critical for success in the international environment—a hybrid culture that bridges the gap between partners in multinational operations. Strategic leaders take time to learn about their partners' cultures—including political, social, and economic aspects. Cultural sensitivity and geopolitical awareness are critical tools for getting things done beyond the traditional chain of command.

12-91. When the Army's immediate needs conflict with the objectives of other agencies, strategic leaders should work to reconcile the differences. Continued disagreement can impair the Army's ability to serve the Nation. Consequently, strategic leaders must devise Army courses of action that reflect national policy objectives and take into account the interests of other organizations and agencies.

Leveraging Technology

12-92. Leveraging superior U.S. technology has given strategic leaders advantages in force projection, command and control, and the generation of overwhelming combat power. Leveraging technology has increased the tempo of operations, the speed of maneuver, the precision of firepower, and the pace at which critical information is processed. Well-managed information technology enhances not only communication, but also situational understanding. Operation Iraqi Freedom in 2003 clearly demonstrated this technological synergism when U.S. forces rapidly defeated Iraq's conventionally employed military.

12-93. Part of leveraging emerging technology includes envisioning desired future capabilities that could be exploited with a particular technology. Another aspect is rethinking the shape and composition of organizations to take advantage of new processes previously not available.

ACCOMPLISHES MISSIONS CONSISTENTLY AND ETHICALLY

12-94. To be able to put strategic vision, concepts, and plans into reality, strategic leaders must employ reliable feedback systems to monitor progress and adherence to values and ethics. They have to find ways to assess many environmental elements to determine the successfulness of policies, operations, or a transformational vision. Like leaders at other levels, they must assess themselves; their leadership style, strengths, and weaknesses; and their fields of excellence. Other assessment efforts involve understanding the will and opinions of the American people, expressed partly through law, policy, their leaders, and the media.

12-95. To gain a complete picture, strategic leaders cast a wide net to assess their own organizations. They develop performance indicators to signal how well they are communicating to all levels of command and how well established systems and processes are balancing the imperatives of doctrine, organization, training, materiel, leadership and education, personnel, and facilities. Assessment starts early in each operation and continues through successful conclusion. They may include monitoring such diverse areas as resource use, development of subordinates, efficiency, effects of stress and fatigue, morale, ethical considerations, and mission accomplishment.

12-96. Strategic leaders routinely deal with diversity, complexity, ambiguity, rapid change, uncertainty, and conflicting policies. They are responsible for developing well-reasoned positions and provide their views and advice to the Nation's highest leaders. For the good of the Army and the Nation, strategic leaders seek to determine what is important now and what will be important in the future.

12-97. General Gordon Sullivan signaled the Army's commitment to learn today's lessons and apply them for visionary concepts of the future when he assigned General Frederick Franks, Jr., V Corps (commander during Operation Desert Storm) as the commanding general of the U.S. Army Training and Doctrine Command. In his guidance to General Franks, General Sullivan specified—

> *You will be informing us and, in turn, teaching us how to think about war in this proclaimed "New World Order," Goldwater-Nichols era in which we are living. What we think about doctrine, organizations, equipment, and training in the future must be the result of a vigorous and informed discussion amongst seasoned professionals.*

12-98. The professionals General Sullivan implied are none other than the competent leaders who serve at all levels of our Army.

Appendix A
Leader Attributes and Core Leader Competencies

A-1. The core leader competencies stem directly from the Army definition of leadership:

> ***Leadership*** is influencing people by providing purpose, motivation, and direction while operating to accomplish the mission and improve the organization.

A-2. The definition contains three basic goals: to lead others, to develop the organization and its individual members, and to accomplish the mission. These goals are extensions of the Army's strategic goal of remaining relevant and ready through effective leadership. The leadership requirements model outlines the attributes and competencies Army leaders develop to meet these goals. (See figure A-1.)

Figure A-1. The Army leadership requirements model

CORE LEADER COMPETENCIES

A-3. The core leader competencies emphasize the roles, functions, and activities of what leaders do. The following discussions and figures provide additional detail on component categories and actions that help convey what each competency involves. The action-based competencies do not include attributes of character (for example, enthusiasm, cooperativeness, flexibility), which are described separately.

LEADS

A-4. Leading is all about influencing others. Leaders and commanders set goals and establish a vision, and then must motivate or influence others to pursue the goals. Leaders influence others in one of two ways. Either the leader and followers communicate directly, or the leader provides an example through everyday actions. The key to effective communication is to come to a common or shared understanding. Leading by example is a powerful way to influence others and is the reason leadership starts with a foundation of the Army Values and the Warrior Ethos. Serving as a role model requires a leader to display character, confidence, and competence to inspire others to succeed. Influencing outside the normal chain of command is a new way to view leadership responsibilities. Leaders have many occasions in joint, interagency, intergovernmental, and multinational situations to lead through diplomacy, negotiation, conflict resolution, and consensus building. To support these functions, leaders need to build trust inside and outside the traditional lines of authority and need to understand their sphere, means, and limits of influence. (Figures A-2 through A-5 identify the first four competencies and associated components and actions.)

Leads Others Leaders motivate, inspire, and influence others to take initiative, work toward a common purpose, accomplish critical tasks, and achieve organizational objectives. Influence is focused on compelling others to go beyond their individual interests and to work for the common good.	
Establishes and imparts clear intent and purpose	• Determines goals or objectives. • Determines the course of action necessary to reach objectives and fulfill mission requirements. • Restates the higher headquarters' mission in terms appropriate to the organization. • Communicates instructions, orders, and directives to subordinates. • Ensures subordinates understand and accept direction. • Empowers and delegates authority to subordinates. • Focuses on the most important aspects of a situation.
Uses appropriate influence techniques to energize others	• Uses techniques ranging from compliance to commitment (pressure, legitimate requests, exchange, personal appeals, collaboration, rational persuasion, apprising, inspiration, participation, and relationship building).
Conveys the significance of the work	• Inspires, encourages, and guides others toward mission accomplishment. • When appropriate, explains how tasks support the mission and how missions support organizational objectives. • Emphasizes the importance of organizational goals.
Maintains and enforces high professional standards	• Reinforces the importance and role of standards. • Performs individual and collective tasks to standard. • Recognizes and takes responsibility for poor performance and addresses it appropriately.
Balances requirements of mission with welfare of followers	• Assesses and routinely monitors the impact of mission fulfillment on mental, physical, and emotional attributes of subordinates. • Monitors morale, physical condition, and safety of subordinates. • Provides appropriate relief when conditions jeopardize success of the mission or present overwhelming risk to personnel.
Creates and promulgates vision of the future	• Interprets data about the future environment, tasks, and missions. • Forecasts probable situations and outcomes and formulates strategies to prepare for them. • Communicates to others a need for greater understanding of the future environment, challenges, and objectives.

Figure A-2. Competency of *leads others* and associated components and actions

Extends Influence Beyond the Chain of Command Leaders need to influence beyond their direct lines of authority and beyond chains of command. This influence may extend to joint, interagency, intergovernmental, multinational, and other groups. In these situations, leaders use indirect means of influence: diplomacy, negotiation, mediation, arbitration, partnering, conflict resolution, consensus building, and coordination.	
Understands sphere of influence, means of influence, and limits of influence	• Assesses situations, missions, and assignments to determine the parties involved in decision making, decision support, and possible interference or resistance.
Builds trust	• Is firm, fair, and respectful to gain trust. • Identifies areas of commonality. • Engages other members in activities and objectives. • Follows through on actions related to expectations of others. • Keeps people informed of actions and results.
Negotiates for understanding, builds consensus, and resolves conflict	• Leverages trust to establish agreements and courses of action. • Clarifies the situation. • Identifies individual and group positions and needs. • Identifies roles and resources. • Facilitates understanding of conflicting positions. • Generates and facilitates generation of possible solutions. • Gains cooperation or support when working with others.
Builds and maintains alliances	• Establishes contact and interacts with others who share common interests, such as development, reaching goals, and giving advice. • Maintains friendships, business associations, interest groups, and support networks. • Influences perceptions about the organization. • Understands the value of and learns from partnerships, associations, and other cooperative alliances.

Figure A-3. Competency of *extends influence beyond the chain of command* and associated components and actions

Leads By Example Leaders constantly serve as role models for others. Leaders will always be viewed as the example, so they must maintain standards and provide examples of effectiveness through all their actions. All Army leaders should model the Army Values. Modeling provides tangible evidence of desired behaviors and reinforces verbal guidance through demonstration of commitment and action.	
Displays character by modeling the Army Values consistently through actions, attitudes, and communications	• Sets the example by displaying high standards of duty performance, personal appearance, military and professional bearing, physical fitness and health, and ethics. • Fosters an ethical climate. • Shows good moral judgment and behavior. • Completes individual and unit tasks to standard, on time, and within the commander's intent. • Is punctual and meets deadlines. • Demonstrates determination, persistence, and patience.
Exemplifies the Warrior Ethos	• Removes or fights through obstacles, difficulties, and hardships to accomplish the mission. • Demonstrates the will to succeed. • Demonstrates physical and emotional courage. • Communicates how the Warrior Ethos is demonstrated.
Demonstrates commitment to the Nation, Army, unit, Soldiers, community, and multinational partners	• Demonstrates enthusiasm for task completion and, if necessary, methods of accomplishing assigned tasks. • Is available to assist peers and subordinates. • Shares hardships with subordinates. • Participates in team tasks and missions without being asked.
Leads with confidence in adverse situations	• Provides leader presence at the right time and place. • Displays self-control, composure, and positive attitude, especially under adverse conditions. • Is resilient. • Remains decisive after discovering a mistake. • Acts in the absence of guidance. • Does not show discouragement when facing setbacks. • Remains positive when the situation becomes confusing or changes. • Encourages subordinates when they show signs of weakness.
Demonstrates technical and tactical knowledge and skills	• Meets mission standards, protects resources, and accomplishes the mission with available resources using technical and tactical skills. • Displays appropriate knowledge of equipment, procedures, and methods.
Understands the importance of conceptual skills and models them to others	• Displays comfort working in open systems. • Makes logical assumptions in the absence of facts. • Identifies critical issues to use as a guide in making decisions and taking advantage of opportunities. • Recognizes and generates innovative solutions. • Relates and compares information from different sources to identify possible cause-and-effect relationships. • Uses sound judgment and logical reasoning.
Seeks and is open to diverse ideas and points of view	• Encourages respectful, honest communications among staff and decision makers. • Explores alternative explanations and approaches for accomplishing tasks. • Reinforces new ideas; demonstrates willingness to consider alternative perspectives to resolve difficult problems. • Uses knowledgeable sources and subject matter experts. • Recognizes and discourages individuals seeking to gain favor from tacit agreement.

Figure A-4. Competency of *leads by example* and associated components and actions

Communicates Leaders communicate effectively by clearly expressing ideas and actively listening to others. By understanding the nature and importance of communication and practicing effective communication techniques, leaders will relate better to others and be able to translate goals into actions. Communication is essential to all other leadership competencies.	
Listens actively	• Listens and watches attentively. • Makes appropriate notes. • Tunes into content, emotion, and urgency. • Uses verbal and nonverbal means to reinforce with the speaker that you are paying attention. • Reflects on new information before expressing views.
Determines information-sharing strategies	• Shares necessary information with others and subordinates. • Protects confidential information. • Coordinates plans with higher, lower, and adjacent individuals and affected organizations. • Keeps higher and lower headquarters, superiors, and subordinates informed.
Employs engaging communication techniques	• States goals to energize others to adopt and act on them. • Speaks enthusiastically and maintains listeners' interest and involvement. • Makes appropriate eye contact when speaking. • Uses gestures that are appropriate but not distracting. • Uses visual aids as needed. • Acts to determine, recognize, and resolve misunderstandings.
Conveys thoughts and ideas to ensure shared understanding	• Expresses thoughts and ideas clearly to individuals and groups. • Uses correct grammar and doctrinally correct phrases. • Recognizes potential miscommunication. • Uses appropriate means for communicating a message. • Communicates clearly and concisely up, down, across, and outside the organization. • Clarifies when there is some question about goals, tasks, plans, performance expectations, and role responsibilities.
Presents recommendations so others understand advantages	• Uses logic and relevant facts in dialogue. • Keeps conversations on track. • Expresses well-thoughtout and well-organized ideas.
Is sensitive to cultural factors in communication	• Maintains awareness of communication customs, expressions, actions, or behaviors. • Demonstrates respect for others.

Figure A-5. Competency of *communicates* and associated components and actions

DEVELOPS

A-5. Developing the organization, the second category, involves three competencies: creating a positive environment in which the organization can flourish, preparing oneself, and developing other leaders. The environment is shaped by leaders taking actions to foster working together, encouraging initiative and personal acknowledgment of responsibility, setting and maintaining realistic expectations, and demonstrating care for people—the number one resource of leaders. Preparing self involves getting set for mission accomplishment, expanding and maintaining knowledge in such dynamic topic areas as cultural and geopolitical affairs, and being self-aware. Developing others is a directed responsibility of commanders. Leaders develop others through coaching, counseling, and mentoring—each with a different set of implied processes. Leaders also build teams and organizations through direct interaction, resource management, and providing for future capabilities. (Figures A-6 through A-8 [pages A-6 through A-8] identify the three developmental competencies and associated components and actions.)

Creates a Positive Environment Leaders have the responsibility to establish and maintain positive expectations and attitudes that produce the setting for healthy relationships and effective work behaviors. Leaders are charged with improving the organization while accomplishing missions. They should leave the organization better that it was when they arrived.	
Fosters teamwork, cohesion, cooperation, and loyalty	• Encourages people to work together effectively. • Promotes teamwork and team achievement to build trust. • Draws attention to the consequences of poor coordination. • Acknowledges and rewards successful team coordination. • Integrates new members into the unit quickly.
Encourages subordinates to exercise initiative, accept responsibility, and take ownership	• Involves others in decisions and keeps them informed of consequences that affect them. • Allocates responsibility for performance. • Guides subordinate leaders in thinking through problems for themselves. • Allocates decision making to the lowest appropriate level. • Acts to expand and enhance subordinate's competence and self-confidence. • Rewards initiative.
Creates a learning environment	• Uses effective assessment and training methods. • Encourages leaders and their subordinates to reach their full potential. • Motivates others to develop themselves. • Expresses the value of interacting with others and seeking counsel. • Stimulates innovative and critical thinking in others. • Seeks new approaches to problems.
Encourages open and candid communications	• Shows others how to accomplish tasks while remaining respectful, resolute, and focused. • Communicates a positive attitude to encourage others and improve morale. • Reinforces the expression of contrary and minority viewpoints. • Displays appropriate reactions to new or conflicting information or opinions. • Guards against groupthink.
Encourages fairness and inclusiveness	• Provides accurate evaluations and assessments. • Supports equal opportunity. • Prevents all forms of harassment. • Encourages learning about and leveraging diversity.
Expresses and demonstrates care for people and their well-being	• Encourages subordinates and peers to express candid opinions. • Ensures that subordinates and their families are provided for, including their health, welfare, and development. • Stands up for subordinates. • Routinely monitors morale and encourages honest feedback.
Anticipates people's on-the-job needs	• Recognizes and monitors subordinate's needs and reactions. • Shows concern for the impact of tasks and missions on subordinate morale.
Sets and maintains high expectations for individuals and teams	• Clearly articulates expectations. • Creates a climate that expects good performance, recognizes superior performance, and does not accept poor performance. • Challenges others to match the leader's example.
Accepts reasonable setbacks and failures	• Communicates the difference between maintaining professional standards and a zero-defects mentality. • Expresses the importance of being competent and motivated but recognizes the occurrence of failure. • Emphasizes learning from one's mistakes.

Figure A-6. Competency of *creates a positive environment* and associated components and actions

Prepares Self Leaders ensure they are prepared to execute their leadership responsibilities fully. They are aware of their limitations and strengths and seek to develop themselves. Leaders maintain physical fitness and mental well-being. They continue to improve the domain knowledge required of their leadership roles and their profession. Only through continuous preparation for missions and other challenges, being aware of self and situations and practicing lifelong learning and development can an individual fulfill the responsibilities of leadership.	
Maintains mental and physical health and well-being	• Recognizes imbalance or inappropriateness of one's own actions. • Removes emotions from decision making. • Applies logic and reason to make decisions or when interacting with emotionally charged individuals. • Recognizes the sources of stress and maintains appropriate levels of challenge to motivate self. • Takes part in regular exercise, leisure activities, and time away from routine work. • Stays focused on life priorities and values.
Maintains self awareness: employs self understanding, and recognizes impact on others	• Evaluates one's strengths and weaknesses. • Learns from mistakes and makes corrections, learns from experience. • Considers feedback on performance, outcomes associated with actions, and actions taken by others to achieve similar goals. • Seeks feedback on how others view one's own actions. • Routinely determines personal goals and makes progress toward them. • Develops capabilities where possible but accepts personal limitations. • Seeks opportunities where capabilities can be used appropriately. • Understands self-motivation under various task conditions.
Evaluates and incorporates feedback from others	• Determines areas in need of development. • Judges self with the help of feedback from others.
Expands knowledge of technical, technological, and tactical areas	• Keeps informed about developments and policy changes inside and outside the organization. • Seeks knowledge of systems, equipment, capabilities, and situations, particularly information technology systems.
Expands conceptual and interpersonal capabilities	• Understands the contribution of concentration, critical thinking (assimilation of information, discriminating relevant cues, question asking), imagination (decentering), and problem solving in different task conditions. • Learns new approaches to problem solving. • Applies lessons learned. • Filters unnecessary information efficiently. • Reserves time for self-development, reflection, and personal growth. • Considers possible motives behind conflicting information.
Analyzes and organizes information to create knowledge	• Reflects on what has been learned and organizes these insights for future application. • Considers source, quality or relevance, and criticality of information to improve understanding. • Identifies reliable sources of data and other resources related to acquiring knowledge. • Sets up systems or procedures to store knowledge for reuse.
Maintains relevant cultural awareness	• Learns about issues of language, values, customary behavior, ideas, beliefs, and patterns of thinking that influence others. • Learns about results of previous encounters when culture plays a role in mission success.
Maintains relevant geopolitical awareness	• Learns about relevant societies outside the United States experiencing unrest. • Recognizes Army influences on other countries, multinational partners, and enemies. • Understands the factors influencing conflict and peacekeeping, peace enforcing, and peacemaking missions.

Figure A-7. Competency of *prepares self* and associated components and actions

Develops Others Leaders encourage and support others to grow as individuals and teams. They facilitate the achievement of organizational goals through assisting others to develop. They prepare others to assume new positions elsewhere in the organization, making the organization more versatile and productive.	
Assesses current developmental needs of others	• Observes and monitors subordinates under different task conditions to establish strengths and weaknesses. • Notes changes in proficiency. • Evaluates subordinates in a fair and consistent manner.
Fosters job development, job challenge, and job enrichment	• Assesses tasks and subordinate motivation to consider methods of improving work assignments, when job enrichment would be useful, methods of cross-training on tasks, and methods of accomplishing missions. • Designs tasks to provide practice in areas of subordinate's weaknesses. • Designs ways to challenge subordinates and improve practice. • Encourages subordinates to improve processes.
Counsels, coaches, and mentors	• Improves subordinate's understanding and proficiency. • Uses experience and knowledge to improve future performance. • Counsels, coaches, and mentors subordinates, subordinate leaders, and others.
Facilitates ongoing development	• Maintains awareness of existing individual and organizational development programs and removes barriers to development. • Supports opportunities for self-development. • Arranges training opportunities as needed that help subordinates improve self-awareness, confidence, and competence.
Supports institutional-based development	• Encourages subordinates to pursue institutional learning opportunities. • Provides information about institutional training and career progression to subordinates. • Maintains resources related to development.
Builds team or group skills and processes	• Presents challenging assignments for team or group interaction. • Provides resources and support. • Sustains and improves the relationships among team or group members. • Provides realistic, mission-oriented training. • Provides feedback on team processes.

Figure A-8. Competency of *develops others* and associated components and actions

ACHIEVES

A-6. Achieving is the third competency goal. Ultimately, leaders exist to accomplish those endeavors that the Army has prescribed for them. Getting results, accomplishing the mission, and fulfilling goals and objectives are all ways to say that leaders exist at the discretion of the organization to achieve something of value. Leaders get results through the influence they provide in direction and priorities. They develop and execute plans and must consistently accomplish goals to a high ethical standard. (Figure A-9 identifies the eighth core leader competency and associated components and actions.)

Gets Results A leader's ultimate purpose is to accomplish organizational results. A leader gets results by providing guidance and managing resources, as well as performing the other leader competencies. This competency is focused on consistent and ethical task accomplishment through supervising, managing, monitoring, and controlling of the work.	
Prioritizes, organizes, and coordinates taskings for teams or other organizational structures/groups.	• Uses planning to ensure each course of action achieves the desired outcome. • Organizes groups and teams to accomplish work. • Plans to ensure that all tasks can be executed in the time available and that tasks depending on other tasks are executed in the correct sequence. • Limits overspecification and micromanagement.
Identifies and accounts for individual and group capabil-ities and commitment to task	• Considers duty positions, capabilities, and developmental needs when assigning tasks. • Conducts initial assessments when beginning a new task or assuming a new position.
Designates, clarifies, and deconflicts roles	• Establishes and employs procedures for monitoring, coordinating, and regulating subordinat's' actions and activities. • Mediates peer conflicts and disagreements.
Identifies, contends for, allocates, and manages resources	• Allocates adequate time for task completion. • Keeps track of people and equipment. • Allocates time to prepare and conduct rehearsals. • Continually seeks improvement in operating efficiency, resource conservation, and fiscal responsibility. • Attracts, recognizes, and retains talent.
Removes work barriers	• Protects organization from unnecessary taskings and distractions. • Recognizes and resolves scheduling conflicts. • Overcomes other obstacles preventing full attention to accomplishing the mission.
Recognizes and rewards good performance	• Recognizes individual and team accomplishments; rewards them appropriately. • Credits subordinates for good performance. • Builds on successes. • Explores new reward systems and understands individual reward motivations.
Seeks, recognizes, and takes advantage of opportunities to improve performance	• Asks incisive questions. • Anticipates needs for action. • Analyzes activities to determine how desired end states are achieved or affected. • Acts to improve the organization's collective performance. • Envisions ways to improve. • Recommends best methods for accomplishing tasks. • Leverages information and communication technology to improve individual and group effectiveness. • Encourages staff to use creativity to solve problems
Makes feedback part of work processes	• Gives and seeks accurate and timely feedback. • Uses feedback to modify duties, tasks, procedures, requirements, and goals when appropriate. • Uses assessment techniques and evaluation tools (such as AARs) to identify lessons learned and facilitate consistent improvement. • Determines the appropriate setting and timing for feedback.
Executes plans to accomplish the mission	• Schedules activities to meet all commitments in critical performance areas. • Notifies peers and subordinates in advance when their support is required. • Keeps track of task assignments and suspenses. • Adjusts assignments, if necessary. • Attends to details.
Identifies and adjusts to external influences on the mission or taskings and organization	• Gathers and analyzes relevant information about changing situations. • Determines causes, effects, and contributing factors of problems. • Considers contingencies and their consequences. • Makes necessary, on-the-spot adjustments.

Figure A-9. Competency of *gets results* and associated components and actions

ATTRIBUTES

A-7. The core leader competencies are complemented by attributes that distinguish high performing leaders of character. Attributes are characteristics that are an inherent part of an individual's total core, physical, and intellectual aspects. Attributes shape how an individual behaves in their environment. Attributes for Army leaders are aligned to identity, presence, and intellectual capacity. (See figures A-10 through A-12.)

A Leader of Character (Identity) Factors internal and central to a leader, that which makes up an individual's core.	
Army Values	• Values are the principles, standards, or qualities considered essential for successful leaders. • Values are fundamental to help people discern right from wrong in any situation. • The Army has set seven values that must be developed in all Army individuals: loyalty, duty, respect, selfless service, honor, integrity, and personal courage.
Empathy	• The propensity to experience something from another person's point of view. • The ability to identify with and enter into another person's feelings and emotions. • The desire to care for and take care of Soldiers and others.
Warrior Ethos	• The shared sentiment internal to Soldiers that represents the spirit of the profession of arms.

Figure A-10. Attributes associated with a leader of character (identity)

A Leader with Presence How a leader is perceived by others based on the leader's outward appearance, demeanor, actions, and words.	
Military bearing	• Possessing a commanding presence. • Projecting a professional image of authority.
Physically fit	• Having sound health, strength, and endurance that support one's emotional health and conceptual abilities under prolonged stress.
Confident	• Projecting self-confidence and certainty in the unit's ability to succeed in whatever it does. • Demonstrating composure and an outward calm through steady control over one's emotions.
Resilient	• Showing a tendency to recover quickly from setbacks, shock, injuries, adversity, and stress while maintaining a mission and organizational focus.

Figure A-11. Attributes associated with a leader with presence

A Leader with Intellectual Capacity The mental resources or tendencies that shape a leaders' conceptual abilities and impact of effectiveness.	
Agility	• Flexibility of mind. • The tendency to anticipate or adapt to uncertain or changing situations; to think through second- and third-order effects when current decisions or actions are not producing the desired effects. • The ability to break out of mental "sets" or habitual thought patterns; to improvise when faced with conceptual impasses. • The ability to quickly apply multiple perspectives and approaches to assessment, conceptualization, and evaluation.
Judgment	• The capacity to assess situations or circumstances shrewdly and to draw sound conclusions. • The tendency to form sound opinions and make sensible decisions and reliable guesses. • The ability to make sound decisions when all facts are not available.
Innovative	• The tendency to introduce new ideas when then opportunity exists or in the face of challenging circumstances. • Creativity in the production of ideas and objects that are both novel or original and worthwhile or appropriate.
Interpersonal tact	• The capacity to understand interactions with others. • Being aware of how others see you and sensing how to interact with them effectively. • Consciousness of character and motives of others and how that affects interacting with them.
Domain knowledge	• Possessing facts, beliefs, and logical assumptions in relevant areas. • Technical knowledge—specialized information associated with a particular function or system. • Tactical knowledge—understanding military tactics related to securing a designated objective through military means. • Joint knowledge—understanding joint organizations, their procedures, and their roles in national defense. • Cultural and geopolitical knowledge—understanding cultural, geographic, and political differences and sensitivities.

Figure A-12. Attributes associated with a leader with intellectual capacity

Appendix B
Counseling

B-1. Counseling is the process used by leaders to review with a subordinate the subordinate's demonstrated performance and potential (Part Three, Chapter 8).

B-2. Counseling is one of the most important leadership development responsibilities for Army leaders. The Army's future and the legacy of today's Army leaders rests on the shoulders of those they help prepare for greater responsibility.

TYPES OF DEVELOPMENTAL COUNSELING

B-3. Developmental counseling is categorized by the purpose of the session. The three major categories of developmental counseling are—

- Event counseling.
- Performance counseling.
- Professional growth counseling.

EVENT COUNSELING

B-4. Event-oriented counseling involves a specific event or situation. It may precede events such as appearing before a promotion board or attending training. It can also follow events such as noteworthy duty performance, a problem with performance or mission accomplishment, or a personal issue. Examples of event-oriented counseling include—

- Instances of superior or substandard performance.
- Reception and integration counseling.
- Crisis counseling.
- Referral counseling.
- Promotion counseling.
- Separation counseling.

Counseling for Specific Instances

B-5. Sometimes counseling is tied to specific instances of superior or substandard duty performance. The leader uses the counseling session to convey to the subordinate whether or not the performance met the standard and what the subordinate did right or wrong. Successful counseling for specific performance occurs as close to the event as possible. Leaders should counsel subordinates for exceptional as well as substandard duty performance. The key is to strike a balance between the two. To maintain an appropriate balance, leaders keep track of counseling for exceptional versus substandard performance.

B-6. Although good leaders attempt to balance their counseling emphasis, leaders should always counsel subordinates who do not meet the standard. If the Soldier or civilian's performance is unsatisfactory because of a lack of knowledge or ability, leader and subordinate can develop a plan for improvement. Corrective training helps ensure that the subordinate knows and consistently achieves the standard.

B-7. When counseling a subordinate for a specific performance, take the following actions:

- Explain the purpose of the counseling—what was expected, and how the subordinate failed to meet the standard.
- Address the specific unacceptable behavior or action—do not attack the person's character.
- Explain the effect of the behavior, action, or performance on the rest of the organization.

- Actively listen to the subordinate's response.
- Remain neutral.
- Teach the subordinate how to meet the standard.
- Be prepared to do some personal counseling, since a failure to meet the standard may be related to or be the result of an unresolved personal problem.
- Explain to the subordinate how an individual development plan will improve performance and identify specific responsibilities in implementing the plan. Continue to assess and follow up on the subordinate's progress. Adjust the plan as necessary.

Reception and Integration Counseling

B-8. Caring and empathic Army leaders should counsel all new team members when they join the organization. Reception and integration counseling serves two important purposes:

- It identifies and helps alleviate any problems or concerns that new members may have, including any issues resulting from the new duty assignment.
- It familiarizes new team members with the organizational standards and how they fit into the team. It clarifies roles and assignments and sends the message that the chain of command cares.

B-9. Reception and integration counseling should among others include the following discussion points:

- Chain of command familiarization.
- Organizational standards.
- Security and safety issues.
- Noncommissioned officer (NCO) support channel (who is in it and how it is used).
- On- and off-duty conduct.
- Personnel/personal affairs/initial and special clothing issue.
- Organizational history, structure, and mission.
- Soldier programs within the organization, such as Soldier of the Month/Quarter/Year, and educational and training opportunities.
- Off limits and danger areas.
- Functions and locations of support activities.
- On- and off-post recreational, educational, cultural, and historical opportunities.
- Foreign nation or host nation orientation.
- Other areas the individual should be aware of as determined by the leader.

Crisis Counseling

B-10. Crisis counseling includes getting a Soldier or employee through a period of shock after receiving negative news, such as the notification of the death of a loved one. It focuses on the subordinate's immediate short-term needs. Leaders may assist the subordinate by listening and providing appropriate assistance. Assisting can also mean referring the subordinate to a support activity or coordinating for external agency support, such as obtaining emergency funding for a flight ticket or putting them in contact with a chaplain.

Referral Counseling

B-11. Referral counseling helps subordinates work through a personal situation. It may or may not follow crisis counseling. Referral counseling aims at preventing a problem from becoming unmanageable if the empathic Army leader succeeds in identifying the problem in time and involves appropriate resources, such as Army Community Services, a chaplain, or an alcohol and drug counselor. (Figure B-4 lists support activities.)

Promotion Counseling

B-12. Army leaders must conduct promotion counseling for all specialists and sergeants who are eligible for advancement without waivers but not recommended for promotion to the next higher grade. Army regulations require that Soldiers within this category receive initial (event-oriented) counseling when they attain full promotion eligibility and then periodic (performance/personal growth) counseling thereafter.

Adverse Separation Counseling

B-13. Adverse separation counseling may involve informing the Soldier of the administrative actions available to the commander in the event substandard performance continues and of the consequences associated with those administrative actions (see AR 635-200).

B-14. Developmental counseling may not apply when an individual has engaged in serious acts of misconduct. In those situations, leaders should refer the matter to the commander and the servicing staff judge advocate. When rehabilitative efforts fail, counseling with a view towards separation is required. It is an administrative prerequisite to many administrative discharges, while sending a final warning to the Soldier: improve performance or face discharge. In many situations, it is advisable to involve the chain of command as soon as it is determined that adverse separation counseling might be required. A unit first sergeant or the commander should inform the Soldier of the notification requirements outlined in AR 635-200.

PERFORMANCE COUNSELING

B-15. During performance counseling, leaders conduct a review of a subordinate's duty performance over a certain period. Simultaneously, leader and subordinate jointly establish performance objectives and standards for the next period. Rather than dwelling on the past, focus on the future: the subordinate's strengths, areas of improvement, and potential.

B-16. Performance counseling is required under the officer, NCO, and Army civilian evaluation reporting systems. The officer evaluation report (OER) (DA Form 67-9) process requires periodic performance counseling as part of the OER Support Form requirements. Mandatory, face-to-face performance counseling between the rater and the rated NCO is required under the noncommissioned officer evaluation reporting system. (See AR 623-3). Performance evaluation for civilian employees also includes both of these requirements.

B-17. Counseling at the beginning of and during the evaluation period ensures the subordinate's personal involvement in the evaluation process. Performance counseling communicates standards and is an opportunity for leaders to establish and clarify the expected values, attributes, and competencies. The OER support form's coverage of leader attributes and competencies is an excellent tool for leader performance counseling. For lieutenants and junior warrant officers, the major performance objectives on the OER Support Form (DA Form 67-9-1) are used as the basis for determining the developmental tasks on the Developmental Support Form (DA Form 67-9-1A). Quarterly face-to-face performance and developmental counseling is required for these junior officers as outlined in AR 623-3. Army leaders ensure that performance objectives and standards are focused and tied to the organization's objectives and the individual's professional development. They should also echo the objectives on the leader's support form as a team member's performance contributes to mission accomplishment.

PROFESSIONAL GROWTH COUNSELING

B-18. Professional growth counseling includes planning for the accomplishment of individual and professional goals. During the counseling, leader and subordinate conduct a review to identify and discuss the subordinate's strengths and weaknesses and to create an individual development plan that builds upon those strengths and compensates for (or eliminates) weaknesses.

B-19. As part of professional growth counseling, leader and subordinate may choose to develop a "pathway to success" with short- and long-term goals and objectives. The discussion of the pathway includes opportunities for civilian or military schooling, future duty assignments, special programs, and reenlistment options. An individual development plan is a requirement for all Soldiers and Army civilians as every person's needs and interests are different.

B-20. Career field counseling is required for lieutenants and captains before they are considered for promotion to major. Raters and senior raters in conjunction with the rated officer need to determine where the officer's skills and talents best fit the needs of the Army. The rated officer's preference and abilities (both performance and intellectual) must be considered. The rater and senior rater should discuss career field designation with the officer prior to making a recommendation on the rated officer's OER.

B-21. While these categories can help organize and focus counseling sessions, they should not be viewed as separate or exhaustive. For example, a counseling session that focuses on resolving a problem may also address improving duty performance. A session focused on performance often includes a discussion on opportunities for professional growth. Regardless of the topic of the counseling session, leaders should follow a basic format to prepare for and conduct it. The Developmental Counseling Form, DA Form 4856, discussed at the end of this appendix provides a useful framework to prepare for almost any type of counseling. Use it to help mentally organize the relevant issues to cover during counseling sessions.

THE LEADER AS A COUNSELOR

B-22. To be effective, developmental counseling must be a shared effort. Leaders assist their subordinates in identifying strengths and weaknesses and creating plans of action. Once an individual development plan is agreed upon, they support their Soldiers and civilians throughout the plan implementation and continued assessment. To achieve success, subordinates must be forthright in their commitment to improve and candid in their own assessments and goal setting.

B-23. Army leaders evaluate Army civilians using procedures prescribed under civilian personnel policies. DA Form 4856 is appropriate to counsel Army civilians on professional growth and career goals. DA Form 4856 is not adequate to address civilian counseling concerning Army civilian misconduct or poor performance. The servicing Civilian Personnel Office can provide guidance for such situations.

B-24. Caring and empathic Army leaders conduct counseling to help subordinates become better team members, maintain or improve performance, and prepare for the future. While it is not easy to address every possible counseling situation, leader self-awareness and an adaptable counseling style focusing on key characteristics will enhance personal effectiveness as a counselor. These key characteristics include—

- **Purpose**: Clearly define the purpose of the counseling.
- **Flexibility**: Fit the counseling style to the character of each subordinate and to the relationship desired.
- **Respect**: View subordinates as unique, complex individuals, each with a distinct set of values, beliefs, and attitudes.
- **Communication**: Establish open, two-way communication with subordinates using spoken language, nonverbal actions, gestures, and body language. Effective counselors listen more than they speak.
- **Support**: Encourage subordinates through actions while guiding them through their problems.

THE QUALITIES OF THE COUNSELOR

B-25. Army leaders must demonstrate certain qualities to be effective counselors. These qualities include respect for subordinates, self-awareness and cultural awareness, empathy, and credibility.

B-26. One challenging aspect of counseling is selecting the proper approach to a specific situation. To counsel effectively, the technique used must fit the situation, leader capabilities, and subordinate expectations. Sometimes, leaders may only need to give information or listen, while in other situations a subordinate's improvement may call for just a brief word of praise. Difficult circumstances may require structured counseling followed by definite actions, such as referrals to outside experts and agencies.

B-27. Self-aware Army leaders consistently develop and improve their own counseling abilities. They do so by studying human behavior, learning the kinds of problems that affect their followers, and developing their interpersonal skills. The techniques needed to provide effective counseling vary from person to person and session to session. However, general skills that leaders will need in almost every situation include active listening, responding, and questioning.

ACTIVE LISTENING

B-28. Active listening helps communicate reception of the subordinate's message verbally and nonverbally. To capture the message fully, leaders listen to what is said and observe the subordinate's manners. Key elements of active listening include—

- **Eye contact**. Maintaining eye contact without staring helps show sincere interest. Occasional breaks of eye contact are normal and acceptable, while excessive breaks, paper shuffling, and clock-watching may be perceived as a lack of interest or concern.
- **Body posture**. Being relaxed and comfortable will help put the subordinate at ease. However, a too-relaxed position or slouching may be interpreted as a lack of interest.
- **Head nods**. Occasionally head nodding indicates paying attention and encourages the subordinate to continue.
- **Facial expressions**. Keep facial expressions natural and relaxed to signal a sincere interest.
- **Verbal expressions**. Refrain from talking too much and avoid interrupting. Let the subordinate do the talking, while keeping the discussion on the counseling subject.

B-29. Active listening implies listening thoughtfully and deliberately to capture the nuances of the subordinate's language. Stay alert for common themes. A subordinate's opening and closing statements as well as recurring references may indicate his priorities. Inconsistencies and gaps may indicate an avoidance of the real issue. Certain inconsistencies may suggest additional questions by the counselor.

B-30. Pay attention to the subordinate's gestures to understand the complete message. By watching the subordinate's actions, leaders identify the emotions behind the words. Not all actions are proof of a subordinate's feelings but they should be considered. Nonverbal indicators of a subordinate's attitude include—

- **Boredom**. Drumming on the table, doodling, clicking a ballpoint pen, or resting the head in the palm of the hand.
- **Self-confidence**. Standing tall, leaning back with hands behind the head, and maintaining steady eye contact.
- **Defensiveness**. Pushing deeply into a chair, glaring at the leader, and making sarcastic comments as well as crossing or folding arms in front of the chest.
- **Frustration**. Rubbing eyes, pulling on an ear, taking short breaths, wringing the hands, or frequently changing total body position.
- **Interest, friendliness, and openness**. Moving toward the leader while sitting.
- **Anxiety**. Sitting on the edge of the chair with arms uncrossed and hands open.

B-31. Leaders consider each indicator carefully. Although each may reveal something about the subordinate, do not judge too quickly. When unsure look for reinforcing indicators or check with the subordinate to understand the behavior, determine what is underlying it, and allow the subordinate to take responsibility.

RESPONDING

B-32. A leader responds verbally and nonverbally to show understanding of the subordinate. Verbal responses consist of summarizing, interpreting, and clarifying the subordinate's message. Nonverbal responses include eye contact and occasional gestures such as a head nod.

QUESTIONING

B-33. Although focused questioning is an important skill, counselors should use it with caution. Too many questions can aggravate the power differential between a leader and a subordinate and place the subordinate in a passive mode. The subordinate may also react to excessive questioning as an intrusion of privacy and become defensive. During a leadership development review, ask questions to obtain information or to get the subordinate to think deeper about a particular situation. Questions should evoke more than a yes or no answer. Well-posed questions deepen understanding, encourage further explanation, and help the subordinate perceive the counseling session as a constructive experience.

COUNSELING ERRORS

B-34. Dominating the counseling by talking too much, giving unnecessary or inappropriate advice, not truly listening, and projecting personal likes, dislikes, biases, and prejudices all interfere with effective counseling. Competent leaders avoid rash judgments, stereotyping, losing emotional control, inflexible counseling methods, or improper follow-up.

B-35. To improve leader counseling skills, follow these general guidelines:

- To help resolve the problem or improve performance, determine the subordinate's role in the situation and what the subordinate has done.
- Draw conclusions based on more factors than the subordinate's statement.
- Try to understand what the subordinate says and feels; listen to what is said and how it is said
- Display empathy when discussing the problem.
- When asking questions, be sure the information is needed.
- Keep the conversation open-ended and avoid interrupting.
- Give the subordinate your full attention.
- Be receptive to the subordinate's emotions, without feeling responsible to save the subordinate from hurting.
- Encourage the subordinate to take the initiative and to speak aloud.
- Avoid interrogating.
- Keep personal experiences out of the counseling session, unless you believe your experiences will really help.
- Listen more and talk less.
- Remain objective.
- Avoid confirming a subordinate's prejudices.
- Help the subordinates help themselves.
- Know what information to keep confidential and what to present to the chain of command, if necessary.

ACCEPTING LIMITATIONS

B-36. Army leaders cannot help everyone in every situation. Recognize personal limitations and seek outside assistance, when required. When necessary, refer a subordinate to the agency more qualified to help.

B-37. The agency list in figure B-1 assists in solving problems. Although it is generally in an individual's best interest to begin by seeking help from their first-line leaders, caring leaders should respect an individual's preference to contact any of these agencies on their own.

Activity	Description
Adjutant General	Provides personnel and administrative services support such as orders, ID cards, retirement assistance, deferments, and in- and out-processing.
American Red Cross	Provides communications support between Soldiers and families and assistance during or after emergency or compassionate situations.
Army Community Service	Assists military families through their information and referral services, budget and indebtedness counseling, household item loan closet, and information about other military posts.
Army Substance Abuse Program	Provides alcohol and drug abuse prevention and control programs.
Better Opportunities for Single Soldiers (BOSS)	Serves as a liaison between installation agencies and single Soldiers.
Army Education Center	Provides services for continuing education and individual learning services support.
Army Emergency Relief	Provides financial assistance and personal budget counseling; coordinates student loans through Army Emergency Relief education loan programs.
Career Counselor	Explains reenlistment options and provides current information on prerequisites for reenlistment and selective reenlistment bonuses.
Chaplain	Provides spiritual and humanitarian counseling to Soldiers and Army civilians.
Claims Section, SJA	Handles claims for and against the government, most often those for the loss and damage of household goods.
Legal Assistance Office	Provides legal information or assistance on matters of contracts, citizenship, adoption, marital problems, taxes, wills, and powers of attorney.
Community Counseling Center	Provides alcohol and drug abuse prevention and control programs for Soldiers.
Community Health Nurse	Provides preventive health care services.
Community Mental Health Service	Provides assistance and counseling for mental health problems.
Employee Assistance Program	Provides health nurse, mental health service, and social work services for Army civilians.
Equal Opportunity Staff Office and Equal Employment Opportunity Office	Provides assistance for matters involving discrimination in race, color, national origin, gender, and religion. Provides information on procedures for initiating complaints and resolving complaints informally.
Family Advocacy Officer	Coordinates programs supporting children and families including abuse and neglect investigation, counseling, and educational programs.
Finance and Accounting Office	Handles inquiries for pay, allowances, and allotments.
Housing Referral Office	Provides assistance with housing on and off post.
Inspector General	Renders assistance to Soldiers and Army civilians. Corrects injustices affecting individuals and eliminates conditions determined to be detrimental to the efficiency, economy, morale, and reputation of the Army. Investigates matters involving fraud, waste, and abuse.
Social Work Office	Provides services dealing with social problems to include crisis intervention, family therapy, marital counseling, and parent or child management assistance.

Figure B-1. Support activities

ADAPTIVE APPROACHES TO COUNSELING

B-38. An effective leader approaches each subordinate as an individual. Different people and different situations require different counseling approaches. Three approaches to counseling include nondirective, directive, and combined (see Part Three, Chapter 8 for more). These approaches differ in specific techniques, but all fit the definition of counseling and contribute to its overall purpose. The major difference between the approaches is the degree to which the subordinate participates and interacts during a counseling session. Figure B-2 identifies the advantages and disadvantages of each approach.

	Advantages	Disadvantages
Nondirective	Encourages maturity. Encourages open communication. Develops personal responsibility.	More time-consuming. Requires greatest counselor skills.
Directive	Quickest method. Good for people who need clear, concise direction. Allows counselors to use their experience.	Does not encourage subordinates to be part of the solution. Tends to treat symptoms, not problems. Tends to discourage subordinates from talking freely. Solution is the counselor's, not the subordinate's.
Combined	Moderately quick. Encourages maturity. Encourages open communication. Allows counselors to use their experience.	May take too much time for some situations.

Figure B-2. Counseling approach summary chart

COUNSELING TECHNIQUES

B-39. The Army leader can select from several techniques when counseling subordinates. These techniques may cause subordinates to change behavior and improve upon their performance. Counseling techniques leaders may explore during the nondirective or combined_approaches include—

- **Suggesting alternatives.** Discuss alternative actions that the subordinate may take. Leader and subordinate together decide which course of action is most appropriate.
- **Recommending.** Recommend one course of action, but leave the decision to accept it to the subordinate.
- **Persuading.** Persuade the subordinate that a given course of action is best, but leave the final decision to the subordinate. Successful persuasion depends on the leader's credibility, the subordinate's willingness to listen, and mutual trust.
- **Advising.** Advise the subordinate that a given course of action is best. This is the strongest form of influence not involving a command.

B-40. Techniques to use during the directive approach to counseling include—

- **Corrective training.** Teach and assist the subordinate in attaining and maintaining the required standard. A subordinate completes corrective training when the standard is consistently attained.
- **Commanding.** Order the subordinate to take a given course of action in clear, precise words. The subordinate understands the order and will face consequences for failing to carry it out.

THE FOUR-STAGE COUNSELING PROCESS

B-41. Effective Army leaders make use of a four-stage counseling process:

- Identify the need for counseling.
- Prepare for counseling.
- Conduct counseling.
- Follow-up.

STAGE 1: IDENTIFY THE NEED FOR COUNSELING

B-42. Usually organizational policies—such as counseling associated with an evaluation or command directed counseling—focus a counseling session. However, leaders may also conduct developmental counseling whenever the need arises for focused, two-way communication aimed at subordinate's development. Developing subordinates consists of observing the subordinate's performance, comparing it to the standard, and then providing feedback to the subordinate in the form of counseling.

STAGE 2: PREPARE FOR COUNSELING

B-43. Successful counseling requires preparation in the following seven areas:

- Select a suitable place.
- Schedule the time.
- Notify the subordinate well in advance.
- Organize information.
- Outline the counseling session components.
- Plan the counseling strategy.
- Establish the right atmosphere.

Select a Suitable Place

B-44. Conduct the counseling in an environment that minimizes interruptions and is free from distracting sights and sounds.

Schedule the Time

B-45. When possible, counsel a subordinate during the duty day. Counseling after duty hours may be rushed or perceived as unfavorable. Select a time free from competition with other activities. Consider that important events occurring after the session could distract a subordinate from concentrating on the counseling. The scheduled time for counseling should also be appropriate for the complexity of the issue at hand. Generally, counseling sessions should last less than an hour.

Notify the Subordinate Well in Advance

B-46. Counseling is a subordinate-centered, two-person effort for which the subordinate must have adequate time to prepare. The person to be counseled should know why, where, and when the counseling takes place. Counseling tied to a specific event should happen as closely to the event as possible. For performance or professional development counseling, subordinates may need at least a week or more to prepare or review specific documents and resources, including evaluation support forms or counseling records.

Organize Information

B-47. The counselor should review all pertinent information, including the purpose of the counseling, facts, and observations about the person to be counseled, identification of possible problems, and main points of discussion. The counselor can outline a possible plan of action with clear obtainable goals as a basis for the final plan development between counselor and the Soldier or civilian.

Outline the Components of the Counseling Session

B-48. Using the available information, determine the focus and specific topics of the counseling session. Note what prompted the counseling requirement, aims, and counselor role. Identify possible key comments and questions to keep the counseling session subordinate-centered and which can help guide the subordinate through the session's stages. As subordinates may be unpredictable during counseling, a written outline can help keep the session on track and enhances the chance for focused success.

Counseling Outline

Type of counseling: Initial NCOER counseling for SFC Taylor, a recently promoted new arrival to the unit.

Place and time: The platoon office, 1500 hours, 9 October.

Time to notify the subordinate: Notify SFC Taylor one week in advance of the counseling session.

Subordinate preparation: Instruct SFC Taylor to put together a list of goals and objectives he would like to complete over the next 90 to 180 days. Review the values, attributes, and competencies of FM 6-22.

Counselor preparation:

Review the NCO Counseling Checklist/Record

Update or review SFC Taylor's duty description and fill out the rating chain and duty description on the working copy of the NCOER.

Review each of the values and responsibilities in Part IV of the NCOER and the values, attributes, and competencies in FM 6-22. Think of how each applies to SFC Taylor's duties as platoon sergeant.

Review the actions necessary for a success or excellence rating in each value and responsibility.

Make notes in blank spaces on relevant parts of the NCOER to assist in counseling.

Role as a counselor: Help SFC Taylor to understand the expectations and standards associated with the platoon sergeant position. Assist SFC Taylor in developing the values, attributes, and competencies that enable him to achieve his performance objectives consistent with those of the platoon and company. Resolve any aspects of the job that SFC Taylor does not clearly understand.

Session outline: Complete an outline following the counseling session components listed in figure B-4 and based on the draft duty description on the NCOER. This should happen two to three days prior to the actual counseling session.

Figure B-3. Example of a counseling outline

Plan the Counseling Strategy

B-49. There are many different approaches to counseling. The directive, nondirective, and combined approaches offer a variety of options that can suit any subordinates and situation (see figure B-3 and Part Three, Chapter 8).

Establish the Right Atmosphere

B-50. The right atmosphere promotes open, two-way communication between a leader and subordinate. To establish a more relaxed atmosphere, offer the subordinate a seat or a cup of coffee. If appropriate, choose to sit in a chair facing the subordinate since a desk can act as a barrier.

B-51. Some situations require more formal settings. During counseling to correct substandard performance, leaders seated behind a desk may direct the subordinate to remain standing. This reinforces the leader's role and authority and underscores the severity of the situation.

Example Counseling Session

Open the Session

- To establish a relaxed environment for an open exchange, explain to SFC Taylor that the more one discusses and comprehends the importance of the Army Values, leader attributes, and competencies, the easier it is to develop and incorporate them for success into an individual leadership style.
- State the purpose of the counseling session and stress that the initial counseling is based on what SFC Taylor needs to do to be a successful platoon sergeant in the unit. Come to an agreement on the duty description and the specific performance requirements. Discuss related values, competencies, and the standards for success. Explain that subsequent counseling will address his developmental needs as well as how well he is meeting the jointly agreed upon performance objectives. Urge a thorough self-assessment during the next quarter to identify his developmental needs.
- Ensure that SFC Taylor knows the rating chain and resolve any questions he has about his duty position and associated responsibilities. Discuss the close team relationship that must exist between a platoon leader and a platoon sergeant, including the importance of honest, two-way communication.

Discuss the Issue

- Jointly review the duty description as spelled out in the NCOER, including all associated responsibilities, such as maintenance, training, and taking care of Soldiers. Relate the responsibilities to leader competencies, attributes, and values. Revise the duty description, if necessary. Highlight areas of special emphasis and additional duties.
- Clearly discuss the meaning of value and responsibility on the NCOER. Discuss the values, attributes, and competencies as outlined in FM 6-22. Ask focused questions to identify if he relates these items to his role as a platoon sergeant.
- Explain to SFC Taylor that the leader's character, presence, and intellect are the basis for competent leadership and that development of the desired leader attributes requires that Army leaders adopt them through consistent self-awareness and lifelong learning. Emphasize that the plan of action to accomplish major performance objectives must encompass the appropriate values, attributes, and competencies. Underscore that the development of the leader's character can never be separate from the overall plan.

Assist in Developing a Plan of Action (During the Counseling Session)

- Ask SFC Taylor to identify tasks that will facilitate the accomplishment of the agreed-upon performance objectives. Describe each by using the values, responsibilities, and competencies found on the NCOER and in FM 6-22.
- Discuss how each value, responsibility, and competency applies to the platoon sergeant position. Discuss specific examples of success and excellence in each

value and responsibility block. Ask SFC Taylor for suggestions to make the goals objective, specific, and measurable.

- Ensure that SFC Taylor leaves the counseling session with at least one example of a success or excellence bullet statement as well as sample bullet statements for each value and responsibility. Discuss SFC Taylor's promotion goals and ask him what he considers his strengths and weaknesses. Obtain copies of the last two master sergeant selection board results and match his goals and objectives.

Close the Session

- Verify SFC Taylor understands the duty description and performance objectives.
- Stress the importance of teamwork and two-way communication.
- Ensure SFC Taylor understands that you expect him to assist in your development as a platoon leader—both of you have the role of teacher and coach.
- Remind SFC Taylor to perform a self-assessment during the next quarter.
- Set a tentative date during the next quarter for the follow-up counseling.

Notes on Strategy

- Offer to answer any questions SFC Taylor may have.
- Expect SFC Taylor to be uncomfortable with the terms and development process and respond in a way that encourages participation throughout the session

Figure B-4. Example of a counseling session

STAGE 3: CONDUCT THE COUNSELING SESSION

B-52. Caring Army leaders use a balanced mix of formal and informal counseling and learn to take advantage of everyday events to provide subordinates with feedback. Counseling opportunities often appear when leaders encounter subordinates in their daily activities in the field, motor pool, barracks, and wherever else Soldiers and civilians perform their duties. Even during ad-hoc counseling, leaders should address the four basic components of a counseling session:

- Opening the session.
- Discussing the issues.
- Developing a plan of action.
- Recording and closing the session.

Open the Session

B-53. In the session opening, the leader counselor states the purpose and establishes a subordinate-centered setting. The counselor establishes an atmosphere of shared purpose by inviting the subordinate to speak. An appropriate purpose statement might be "SFC Taylor, the purpose of this counseling is to discuss your duty performance over the past month and to create a plan to enhance performance and attain performance goals." If applicable, start the counseling session by reviewing the status of the current plan of action.

Discuss the Issues

B-54. Leader and counseled individual should attempt to develop a mutual and clear understanding of the counseling issues. Use active listening and invite the subordinate to do most of the talking. Respond and ask questions without dominating the conversation but help the subordinate better understand the subject of the counseling session: duty performance, a problem situation and its impact, or potential areas for growth.

B-55. To reduce the perception of bias or early judgment, both leader and subordinate should provide examples or cite specific observations. When the issue is substandard performance, the leader must be clear why the performance did not meet the standard. During the discussion, the leader must clearly establish what the subordinate must do to meet the standard in the future. It is very important that the leader frames the issue at hand as substandard performance and prevents the subordinate from labeling the issue as an unreasonable standard. An exception would be when the leader considers the current standard as negotiable or is willing to alter the conditions under which the subordinate can meet the standard.

Develop a Plan of Action

B-56. A plan of action identifies a method and pathway for achieving a desired result. It specifies what the subordinate must do to reach agreed-upon goals set during the counseling session. The plan of action must be specific, showing the subordinate how to modify or maintain his or her behavior. Example: "PFC Miller, next week you'll attend the map reading class with 1st Platoon. After the class, SGT Dixon will personally coach you through the land navigation course. He will help you develop your skills with the compass. After observing you going through the course with SGT Dixon, I will meet with you again to determine if you still need additional training."

Record and Close the Session

B-57. Although requirements to record counseling sessions vary, a leader always benefits from documenting the main points of a counseling session, even the informal ones. Documentation serves as a ready reference for the agreed-upon plan of action and helps the leader track the subordinate's accomplishments, improvements, personal preferences, or problems. A good record of counseling enables the leader to make proper recommendations for professional development, schools, promotions, and evaluation reports.

B-58. Army regulations require specific written records of counseling for certain personnel actions, such as barring a Soldier from reenlisting, processing an administrative separation, or placing a Soldier in the overweight program. When a Soldier faces involuntary separation, the leader must maintain accurate counseling records. Documentation of substandard actions often conveys a strong message to subordinates that a further slip in performance or discipline could require more severe action or punishment.

B-59. When closing the counseling session, summarize the key points and ask if the subordinate understands and agrees with the proposed plan of action. With the subordinate present, establish any follow-up measures necessary to support the successful implementation of the plan of action. Follow-up measures may include providing the subordinate with specific resources and time, periodic assessments of the plan, and additional referrals. If possible, schedule future meetings before dismissing the subordinate.

STAGE 4: FOLLOW-UP

Leader Responsibilities

B-60. The counseling process does not end with the initial counseling session. It continues throughout the implementation of the plan of action, consistent with the observed results. Sometimes, the initial plan of action will require modification to meet its goals. Leaders must consistently support their subordinates in implementing the plan of action by teaching, coaching, mentoring, or providing additional time, referrals, and other appropriate resources. Additional measures may include more focused follow-up counseling, informing the chain of command, and taking more severe corrective measures.

Assess the Plan of Action

B-61. During assessment, the leader and the subordinate jointly determine if the desired results were achieved. They should determine the date for their initial assessment during the initial counseling session. The plan of action assessment provides useful information for future follow-up counseling sessions.

SUMMARY—THE COUNSELING PROCESS AT A GLANCE

B-62. Use figure B-5 as a quick reference whenever counseling Soldiers or civilian team members.

Leaders must demonstrate these qualities to counsel effectively:	The Counseling Process:
Leaders must demonstrate these qualities to counsel effectively: • Respect for subordinates. • Self and cultural awareness. • Credibility. • Empathy. Leaders must possess these counseling skills: • Active listening. • Responding. • Questioning. Effective leaders avoid common counseling mistakes. Leaders should avoid— • Personal bias. • Rash judgments. • Stereotyping. • Losing emotional control. • Inflexible counseling methods. • Improper follow-up.	The Counseling Process: Identify the need for counseling. Prepare for counseling: • Select a suitable place. • Schedule the time. • Notify the subordinate well in advance. • Organize information. • Outline the components of the counseling session. • Plan counseling strategy. • Establish the right atmosphere. Conduct the counseling session: • Open the session. • Discuss the issue. • Develop a plan of action (to include the leader's responsibilities). • Record and close the session. Follow up: • Support plan of action implementation. • Assess the plan of action.

Figure B-5. A summary of counseling

THE DEVELOPMENTAL COUNSELING FORM

B-63. The Developmental Counseling Form (DA Form 4856) is designed to help Army leaders conduct and record counseling sessions. Figures B-6 and B-7 show a completed DA Form 4856 documenting the counseling of a young Soldier with financial problems. Although derogatory, it is still developmental counseling. Leaders must decide when counseling, additional training, rehabilitation, reassignment, or other developmental options have been exhausted. Figures B-8 and B-9 show a routine performance/professional growth counseling for a unit first sergeant. Figures B-10 and B-11 show a blank form with instructions on how to complete each block.

DEVELOPMENTAL COUNSELING FORM

For use of this form see FM 6-22; the proponent agency is TRADOC

DATA REQUIRED BY THE PRIVACY ACT OF 1974

AUTHORITY: 5 USC 301, Departmental Regulations; 10 USC 3013, Secretary of the Army and E.O. 9397 (SSN)
PRINCIPAL PURPOSE: To assist leaders in conducting and recording counseling dat a pertaining to subordinates.
ROUTINE USES: For subordinate leader development IAW FM 6-22. Leaders should use this form as necessary.
DISCLOSURE: Disclosure is voluntary.

PART I - ADMINISTRATIVE DATA

Name *(Last, First, MI)*	Rank / Grade	Social Security No.	Date of Counseling
Jones, Andrew	PFC	123-45-6789	28 April 2006

Organization	Name and Title of Counselor
2nd Platoon, B Battery, 1-1 ADA BN	SGT Mark Levy, Squad Leader

PART II - BACKGROUND INFORMATION

Purpose of Counseling: *(Leader states the reason for the counseling, e.g. Performance/Professional or Event-Oriented counseling and includes the leaders facts and observations prior to the counseling):*

To inform PFC Jones of his responsibility to manage his financial affairs and the potential consequence of poor management. To help PFC Jones develop a plan of action to resolve his financial problems.
Facts: The battery commander received reports from the Enlisted Club that PFC Jones had checks returned for insufficient funds. The Enlisted Club cashier has 2 checks for a total of $200 that were returned by American Bank, NA.
A total of $240 is due to the club system for the amount of the checks and fees.

PART III - SUMMARY OF COUNSELING
Complete this section during or immediately subsequent to counseling.

Key Points of Discussion:
PFC Jones, late payments and bounced checks reflect a lack of responsibility and poor management of financial assets. You should know that passing bad checks is a punishable offence under the UCMJ and local law. The commander has been contacted and has the attention of the battery chain of command. The commander, first sergeant and platoon sergeant have begun to question your ability to manage your personal affairs. I also want to remind you that promotions and awards are based on more than just MOS related duties; Soldiers must act responsibly and professionally in all areas of their lives.

Per conversation with PFC Jones, the following information was obtained:
PFC Jones had cashed the checks to purchase food, pay his phone bill and send money home to assist his grandmother with her heating bills. PFC Jones stated he had miscalculated the amount of money in his checking account and will not be able to cover the checks until he gets paid at the end of April 2006. He also stated that warmer weather will reduce any further need to help with his grandmother's utilities. PFC Jones and I went to Army Community Services and they determined the following:

PFC Jones monthly obligations:
Car payment: $330, Car insurance: $138, Rent and utilities: $400. Other credit cards/accounts: $0 Monthly net pay: $1232.63

We discussed that the remaining $364 should cover PFC Jones monthly living expenses. We also discussed that PFC Jones should start a savings account to draw from in emergencies. Although it is not wrong for him to help his grandmother, he needs to make sure that he is not putting his financial stability in jeopardy. He confirmed he wants to get his finances back on track and begin to put money aside in a savings account to prepare for future needs.

OTHER INSTRUCTIONS
This form will be destroyed upon: reassignment *(other than rehabilitative transfers)*, separation at ETS, or upon retirement. For separation requirements and notification of loss of benefits/consequences see local directives and AR 635-200.

DA FORM 4856, MAR 2006 EDITION OF JUN 99 IS OBSOLETE

Figure B-6. Example of a developmental counseling form—event counseling

Plan of Action*: (Outlines actions that the subordinate will do after the counseling session to reach the agreed upon goal(s). The actions must be specific enough to modify or maintain the subordinate's behavior and include a specific time line for implementation and assessment (Part IV below):*

Based on our discussion, PFC Jones will be able to repay the dishonored checks at the Enlisted Club at the end of the month. In the future he will think through his decisions related to his economic needs. PFC Jones has contacted to Enlisted Club and the manager has agreed to give him until 2 May 2006 to redeem the checks. In the future he plans to put money in savings to assist his grandmother if the need arises. His long-term goal is to start a savings account and deposit $50 a month.

PFC Jones is also enrolled in the ACS check cashing and money management classes scheduled for 2 and 9 May 2006.

Assessment Date: 28 July 2006

Session Closing*: (The leader summarizes the key points of the session and checks if the subordinate understands the plan of action. The subordinate agrees/disagrees and provides remarks if appropriate):*

Individual counseled: ☐ I agree ☐ disagree with the information above

Individual counseled remarks:

Signature of Individual Counseled: Andrew Jones Date: 28 April 2006

Leader Responsibilities*: (Leader's responsibilities in implementing the plan of action):*

PFC Jones will visit the manager of the Enlisted Club and repay the $240 for his bad checks. He will provide me a receipt showing the bill has been paid in full. PFC Jones will also provide me with a copy of his budget that ACS will help him develop during his financial management classes.

PFC Jones financial situation will be a key topic in his May 2006 monthly performance counseling session.

Signature of Counselor: Mark Levy Date: 28 April 2006

PART IV - ASSESSMENT OF THE PLAN OF ACTION

Assessment*: (Did the plan of action achieve the desired results? This section is completed by both the leader and the individual counseled and provides useful information for follow-up counseling):*

To be completed during the assessment date in the plan of action.

Counselor: ________________ Individual Counseled: ______________ Date of Assessment: ______________

Note: Both the counselor and the individual counseled should retain a record of the counseling.

REVERSE, DA FORM 4856, MAR 2006

Figure B-7. Example of a developmental counseling form—event counseling (reverse)

DEVELOPMENTAL COUNSELING FORM

For use of this form see FM 6-22; the proponent agency is TRADOC

DATA REQUIRED BY THE PRIVACY ACT OF 1974

AUTHORITY: 5 USC 301, Departmental Regulations; 10 USC 3013, Secretary of the Army and E.O. 9397 (SSN)
PRINCIPAL PURPOSE: To assist leaders in conducting and recording counseling dat a pertaining to subordinates.
ROUTINE USES: For subordinate leader development IAW FM 6-22. Leaders should use this form as necessary.
DISCLOSURE: Disclosure is voluntary.

PART I - ADMINISTRATIVE DATA

Name *(Last, First, MI)*	Rank / Grade	Social Security No.	Date of Counseling
Donalo, Steven	1SG	333-33-3333	12 June 2006

Organization	Name and Title of Counselor
D Company, 3-95th IN BN	CPT Ralph Pedersen, Company Commander

PART II - BACKGROUND INFORMATION

Purpose of Counseling: *(Leader states the reason for the counseling, e.g. Performance/Professional or Event-Oriented counseling and includes the leaders facts and observations prior to the counseling):*

To discuss duty performance for the period 9 March 2006 to 12 June 2006.
To discuss short-range professional growth/plan for next year.
Talk about long-range professional growth (2-5 years) goals.

PART III - SUMMARY OF COUNSELING
Complete this section during or immediately subsequent to counseling.

Key Points of Discussion:

Performance (sustain):

- Emphasized safety, knowledge of demolitions, and tactical proficiency on the Platoon Live Fire Exercises.
- Took charge of company defense during the last major field training exercise; outstanding integration and use of engineers, heavy weapons, and air defense artillery assets in a combined arms environment. Superb defense preparation and execution.
- No dropped white-cycle taskings.
- Good job coordinating with the battalion adjutant on legal and personnel issues.
- Continue to take care of Soldiers; keep the commander abreast of problems.
- Focused on subordinate NCO development; putting the right NCO in the right job.

Improve:
- Get NCOPDs on the calendar
- Hold NCOs to standard on sergeants' time training.

OTHER INSTRUCTIONS

This form will be destroyed upon: reassignment *(other than rehabilitative transfers)*, separation at ETS, or upon retirement. For separation requirements and notification of loss of benefits/consequences see local directives and AR 635-200.

DA FORM 4856, MAR 2006 EDITION OF JUN 99 IS OBSOLETE

Figure B-8. Example of a developmental counseling form—performance/professional growth counseling

Plan of Action*: (Outlines actions that the subordinate will do after the counseling session to reach the agreed upon goal(s). The actions must be specific enough to modify or maintain the subordinate's behavior and include a specific time line for implementation and assessment (Part IV below):*

Developmental Plan (next year):
- Develop a year-long plan for NCOPDs; place on the calendar and training schedules.
- Resume civilian education and correspondence courses.
- Develop a company Soldier of the month competition.
- Assist the company XO in the re-design of the supply room to improve efficiency of EDRE load-outs.
- Put in place a program to develop Ranger School Candidates

Long Range goals (2-5 years):
-Complete Bachelor's degree program
- Attend Sergeant Majors' Academy

Session Closing*: (The leader summarizes the key points of the session and checks if the subordinate understands the plan of action. The subordinate agrees/disagrees and provides remarks if appropriate):*
Individual counseled: ☐ I agree ☐ disagree with the information above
Individual counseled remarks:

Signature of Individual Counseled: Steven Donalo Date: 12 June 2006

Leader Responsibilities*: (Leader's responsibilities in implementing the plan of action):*

Signature of Counselor: Ralph Pedersen Date: 12 June 2006

PART IV - ASSESSMENT OF THE PLAN OF ACTION

Assessment*: (Did the plan of action achieve the desired results? This section is completed by both the leader and the individual counseled and provides useful information for follow-up counseling):*

1SG Donalo has enrolled in an associates' degree program at Webster University. The supply room received all GOs on the latest Command Inspection. Five of sever Ranger applicants successfully completed Ranger School, exceeding the overall course completion rate of 39%. Current OPTEMPO has prevented starting a Soldier of the Month board but the company does hold quarterly boards during the white cycle. The Brigade Command Sergeant Major recently commented on the quality of instruction and planning for the last company NCOPD and presented the NCO instructor with a brigade coin.

Counselor: ______________ Individual Counseled: ____________ Date of Assessment: ____________

Note: Both the counselor and the individual counseled should retain a record of the counseling.

REVERSE, DA FORM 4856, MAR 2006

Figure B-9. Example of a developmental counseling form—performance/professional growth counseling (reverse)

DEVELOPMENTAL COUNSELING FORM

For use of this form see FM 6-22; the proponent agency is TRADOC

DATA REQUIRED BY THE PRIVACY ACT OF 1974

AUTHORITY:	5 USC 301, Departmental Regulations; 10 USC 3013, Secretary of the Army and E.O. 9397 (SSN)
PRINCIPAL PURPOSE:	To assist leaders in conducting and recording counseling data pertaining to subordinates.
ROUTINE USES:	For subordinate leader development IAW FM 6-22. Leaders should use this form as necessary.
DISCLOSURE:	Disclosure is voluntary.

PART I - ADMINISTRATIVE DATA

Name *(Last, First, MI)*	Rank / Grade	Social Security No.	Date of Counseling
Organization		Name and Title of Counselor	

PART II - BACKGROUND INFORMATION

Purpose of Counseling: *(Leader states the reason for the counseling, e.g. Performance/Professional or Event-Oriented counseling and includes the leaders facts and observations prior to the counseling):*

See Paragraph B-53 Open the Session

The leader should annotate pertinent, specific, and objective facts and observations made. If applicable, the leader and subordinate start the counseling session by reviewing the status of the previous plan of action.

PART III - SUMMARY OF COUNSELING

Complete this section during or immediately subsequent to counseling.

Key Points of Discussion:

See paragraph B-54 and B-55 Discuss the Issues.

The leader and subordinate should attempt to develop a mutual understanding of the issues. Both the leader and the subordinate should provide examples or cite specific observations to reduce the perception that either is unnecessarily biased or judgmental.

OTHER INSTRUCTIONS

This form will be destroyed upon: reassignment *(other than rehabilitative transfers)*, separation at ETS, or upon retirement. For separation requirements and notification of loss of benefits/consequences see local directives and AR 635-200.

DA FORM 4856, MAR 2006 EDITION OF JUN 99 IS OBSOLETE

Figure B-10. Guidelines on completing a developmental counseling form

Plan of Action*: (Outlines actions that the subordinate will do after the counseling session to reach the agreed upon goal(s). The actions must be specific enough to modify or maintain the subordinate's behavior and include a specific time line for implementation and assessment (Part IV below):*

See paragraph B-56 Develop a Plan of Action

The plan of action specifies what the subordinate must do to reach the goals set during the counseling session. The plan of action must be specific and should contain the outline, guideline(s), and time line that the subordinate follows. A specific and achievable plan of action sets the stage for successful subordinate development.

Session Closing*: (The leader summarizes the key points of the session and checks if the subordinate understands the plan of action. The subordinate agrees/disagrees and provides remarks if appropriate):*

Individual counseled: ☐ I agree ☐ disagree with the information above

Individual counseled remarks:

See paragraph B-57 through B-59 Close the Session

Signature of Individual Counseled: ______________________ Date:____________

Leader Responsibilities*: (Leader's responsibilities in implementing the plan of action):*

See paragraph B-60 Leader's Responsibilities

To accomplish the plan of action, the leader must list the resources necessary and commit to providing them to the Soldier.

Signature of Counselor: ______________________ Date: ____________

PART IV - ASSESSMENT OF THE PLAN OF ACTION

Assessment*: (Did the plan of action achieve the desired results? This section is completed by both the leader and the individual counseled and provides useful information for follow-up counseling):*

See paragraph B-61 Assess the Plan of Action

The assessment of the plan of action provides useful information for future follow-up counseling. This block should be completed prior to the start of a follow-up counseling session. During an event-oriented counseling session, the counseling session is not complete until this block is completed.

During performance/professional growth counseling, this block serves as the starting point for future counseling sessions. Leaders must remember to conduct this assessment based on resolution of the situation or the established time line discussed in the plan of action block above.

Counselor: ____________ Individual Counseled:____________ Date of Assessment: ____________

Note: Both the counselor and the individual counseled should retain a record of the counseling.

REVERSE, DA FORM 4856, MAR 2006

Figure B-11. Guidelines on completing a developmental counseling form (reverse)

Source Notes

These are sources quoted or paraphrased in this publication. They are listed by page number. Quotations are identified by the first few word of the quote. Where a quote is embedded within a paragraph, the paragraph number is listed. Boldface indicates the title of historical vignettes.

PART ONE THE BASIS OF LEADERSHIP

Chapter 2: The Foundations of Army Leadership

2-1 “When we assumed…”: Trevor Royle, *A Dictionary of Military Quotations* (New York: Simon and Schuster, 1989), 63 (hereafter referred to as Royle).

2-2 Oath of Enlistment: DD Form 4, *Enlistment/Reenlistment Document Armed Forces of the United States*, 10 USC 502.

2-2 Oath of Office: DA Form 71, *Oath of Office–Military Personnel*; 5 USC 3331. The oath administered to commissioned officers includes the words, “*I [full name], having been appointed a [rank] in the United States Army....*”

2-3 “When you are commanding…”: H. A. DeWeerd, ed., *Selected Speeches and Statements of General of the Army George C. Marshall* (Washington, DC: Infantry Journal Press, 1945), 176.

2-3 “Just as the diamond…”: *The Chiefs of Staff, United States Army: On Leadership and the Profession of Arms* (Washington, DC: The Information Management Support Center, 24 March 1997), 10 (hereafter referred to as *Chiefs of Staff* 1997).

2-5 **Colonel Chamberlain at Gettysburg**: John J. Pullen, *The Twentieth Maine* (1957; reprint, Dayton, OH: Press of Morningside Bookshop, 1980), 114-125.

Chapter 3: Leadership Roles, Leadership Levels, and Leader Teams

3-2 3-11: Incorporates 10 USC 3583 *Requirement of Exemplary Service and the Army Values.*

3-3 NCO Vision: *The Army Noncommissioned Officer Guide*, FM 7-22.7 (Headquarters, Department of the Army, 2002) (hereafter referred to as *NCO Guide*).

3-4 3-22: James B. Gunlicks, Acting Director of Training, SUBJECT: “Army Training and Leader Development Panel–Civilian (ATLDP-CIV), Implementation Process Action Team (IPAT) Implementation Plan–ACTION MEMORANDUM,” memorandum for Chief of Staff, Army, 28 May 2003.

3-5 “NCOs like to make…”: Dennis Steele, “Broadening the Picture Calls for Turning Leadership Styles,” *Army Magazine* 39, no. 12 (December 1989): 39.

3-9 3-54 Team Structures. Frederic J. Brown, “Vertical Command Teams,” IDA Document D-2728 (Alexandria, VA: Institute for Defense Analyses, 2002), l-1.

3-10 **Shared Leadership Solves Logistics Challenges**: John Pike, “Operation Enduring Freedom-Afghanistan,” Global Security Web site (7 March 2005): <http://www.globalsecurity.org/military/ops/enduring-freedom.htm>.

3-12 **Stepping Up to Lead***:* Ann Scott Tyson, “Anaconda: A War Story,” *Christian Science Monitor* (1 August 2002): <http://www.csmonitor.com/2002/0801/p01s03-wosc.htm>. Mark Thompson, “Randal Perez Didn’t Join the Army to Be a Hero,” *Time Magazine* (1 September 2002): <http://www.time.com/time/covers/1101020909/aperez.html>. U. S. Department of Defense, “Interview with U.S. Army Soldiers who Participated in Operation Anaconda,” United States Department of Defense Web site (7 March 2002): <http://www.defenselink.mil/Transcripts/Transcript.aspx?TranscriptID=2914>.

PART TWO THE ARMY LEADER: PERSON OF CHARACTER, PRESENCE, AND INTELLECT

Chapter 4: Leader Character

4-1 “Just as fire tempers…”: Margaret Chase Smith, speech to graduating women naval officers at Naval Station, Newport, RI (Skowhegan, ME: Margaret Chase Smith Library, 1952).

4-2 **Soldier Shows Character and Discipline**: TRADOC Pam 525-100-4, *Leadership and Command on the Battlefield: Noncommissioned Officer Corps* (Fort Monroe, VA, 1994), 26.

4-3 “Loyalty is the big…”: S. L. A. Marshall, *Men Against Fire: The Problem of Battle Command in Future War* (Gloucester, MA: Peter Smith, 1978), 200.

4-3 **Loyal in War and Captivity**: Ronald H. Spector, *Eagle Against the Sun* (New York: Random House, 1985). A. J. P. Taylor and S. L. Mayer, *History of World War II* (London: Octopus Books, 1974), 98-111. Department of Veterans Affairs: casualty numbers.

4-4 4-10 “There is a great deal…”: George S. Patton, Jr., *War as I Knew It* (Boston: Houghton Mifflin Company, 1975), 366 (hereafter referred to as Patton).

4-5 “I go anywhere…”: James H. Webb, *A Country Such as This* (Annapolis, MD: Naval Institute Press, March 2001), 247.

4-5 “The discipline which …”: John M. Schofield, *Manual for Noncommissioned Officers and Privates of Infantry of the Army of the United States* (West Point, NY: U.S. Military Academy Library Special Collections, 1917), 12.

4-6 “… [A]sk not what…”: John Bartlett, ed., *Familiar Quotations: A Collection of Passages, Phrases and Proverbs Traced to Their Sources in Ancient and Modern Literature* (Boston: Little, Brown and Company, 1968), 1073.

4-6 “War must be…”: William T. Coffey, Jr., *Patriot Hearts: An Anthology of American Patriotism* (Colorado Springs, CO: Purple Mountain Publishing, 2000), 360 (hereafter referred to as Coffey).

4-7 **Honor, Courage and Selfless Service in Korea: “**Medic on a Mission: An Army Medics Strong-Arm Tactics Help to Carry the Day,” Medal of Honor Web site: <http://www.medalofhonor.com/DavidBleak.htm>.

4-7 “No nation…”: Coffey, 95.

4-8 “Courage is doing…”: Coffey, 123.

4-8 **Courage and Inspiration for Soldiers Then and Now**: “African-American Vet Receives Medal of Honor,” Home of Heroes Web site: <http://www.homeofheroes.com/news/archives> scroll to 1997, select Jan 13. S. H. Kelly, “Seven WWII Vets to Receive Medals of Honor,” *Army News Service* (13 January 1997): <http://www4.army.mil/ocpa/print.php?story_id_key=2187>. “The Only Living African American World War II Hero to Receive the Medal of Honor,” Medal of Honor Web site (13 January 1997): <http://www.medalofhonor.com/VernonBaker.htm>. “Seven Black Soldiers from WWII Tapped to Receive Medal of Honor,” *Boston Sunday Globe* (28 April 1996): <http://www.366th.org/960428.htm>.

4-9 4-39 “Our landings…”: Harry C. Butcher, *My Three Years with Eisenhower: The Personal Diary of Captain Harry C. Butcher, USNR, Naval Aide to General Eisenhower, 1942 to 1945* (New York: Simon and Schuster, 1946), 610.

4-9 4-40 “The concept of…”: William Connelly, “NCOs: It’s Time to Get Tough,” *Army Magazine* 31, no. 10 (October, 1981): 31.

4-10 4-46 “Every organization has…”: Eric K. Shinseki, General, United States Army, SUBJECT: “Implementing Warrior Ethos for the Army,” memorandum for Vice Chief of Staff, Army, 3 June 2005.

4-11 “Wars may be…”: Royle, 48.

4-12 **Task Force Kingston**: Martin Blumenson, “Task Force Kingston,” *Army Magazine* (April 1964): 50-60.

4-14 **He Never Gave In:** Linda Busetti, “Local Vietnam War Hero Receives Medal of Honor,” *Arlington Catholic Herald* (11 July 2002). “MOH Citation for Humbert Roque Versace,” Home of Heroes Web site: <http://www.homeofheroes.com/moh>, scroll to bottom of web page, click on Search Our Site, type in *Versace* in Google search, select MOH Citation for Humbert Roque Versace. Arlington National Cemetery, “Remains Never Recovered,” Arlington National Cemetery Web site: <http://www.arlingtoncemetery.net/medalofh.htm>, scroll to Humbert Roque Versace, USA. “Capt. Humbert Roque “Rocky” Versace, Captured by Viet Cong in 1963 and Executed in 1965,” Special Operations Memorial Web site: <http://www.somf.org/moh/>, scroll to Versace, H.R. “Rocky” USA.

4-15 **Warrant Officer Thompson at My Lai**: James S. Olson and Randy Roberts, *My Lai: A Brief History with Documents* (Boston: Bedford Books, 1998), 159, 909-92. W. R. Peters, *The My Lai Inquiry* (New York: W.W. Norton, 1979), 66-76.

4-15 4-68 Joseph and Edna Josephson Institute of Ethics, *Making Ethical Decisions,* Joseph and Edna Josephson Institute of Ethics Web site: <http://www.josephsoninstitute.org/MED/MED-intro+toc.htm>.

Chapter 5: Leader Presence

5-1 "...[L]eadership is...": *The Noncom's Guide: An Encyclopedia of Information for All Noncommissioned Officers of the United States Army* 16th ed. (Harrisburg: The Stackpole Company, 1962), 38.

5-1 "Our quality soldiers...": Julius W. Gates, "From the Top," *Army Trainer* 9, no. 1 (Fall 1989): 5.

5-2 "...I am obliged...": Burke Davis, *They Called Him Stonewall* (New York: Rinehart & Company, Inc., 1954), 50.

5-3 **Mission First–Never Quit!:** "Army Awards MPs for Turning Table on Ambush," *Army News Service* (16 June 2005): <http://www4.army.mil/ocpa/read.php?story_id_key=7472>. Robin Burk, "Tell Me Again About Women in Combat," Winds of Change Web site (25 March 2005): <http://www.windsofchange.net/archives/006564.php>. SGT Sara Ann Wood, "Female Soldier Receives Silver Star in Iraq," *American Forces Press Service* (17 June 2005): <http://www4.army.mil/ocpa/print.php?story_id_key=7474>. Dogen Hannah, "In Iraq, U.S. Military Women Aren't Strangers to Combat," *Knight Ridder Newspapers* (7 April 2005). Ann Scott Tyson, "Soldier Earns Silver Star for Her Role in Defeating Ambush," *Washington Post* (17 June 2005): A-21.

Chapter 6: Leader Intelligence

6-1 "It's not genius...": Robert Debs Heinl, Jr., *Dictionary of Military and Naval Quotations* (Annapolis, MD: U.S. Naval Institute Press, 1966), 239.

6-2 "Judgment comes...": Omar N. Bradley, "Leadership: An Address to the U.S. Army War College, 07 October 1971," *Parameters* 1(3) (1972): 8.

6-3 "...[A]n officer...": James E. Moss, *Noncommissioned Officers' Manual* (Menasha, WI: George Banta Publishing Co., 1917), 33.

6-4 **Self-Control**: Francis Hesselbein, ed., *Leader to Leader* (New York: Leader to Leader Institute, 2005), 28–29.

6-4 "...[A]nyone can get angry...": Aristotle, *Nicomachean Ethics,* trans. Martin Ostwald (New York: Macmillan Publishing Co., 1962), 50.

6-5 "The commander...": Royle, 173.

6-7 "If you can...": Royle, 37.

6-8 **No Slack Soldiers Take a Knee**: Julian E. Barnes, "A Thunder Run Up Main Street," *U.S. News and World Report* (14 April 2003): <http://www.usnews.com/usnews/news/articles/030414/14front_2.htm>. Edwin Black, *Banking on Baghdad: Inside Iraq's 7,000-Year History of War, Profit, and Conflict* (New York: John Wiley and Sons, 2004). Jim Lacey, "From the Battlefield," *Time Magazine* (14 April 2003): <http://www.time.com/time/magazine/article/0,9171,1004638,00.html>.

PART THREE COMPETENCY-BASED LEADERSHIP FOR DIRECT THROUGH STRATEGIC LEVELS

Chapter 7: Leading

7-1 "In short, Army...": Bob Kerr, "CGSC Class of 2005 Graduates," *Fort Leavenworth Lamp* (23 June 2005): 12.

7-3 "The American soldier...": Omar N. Bradley, "American Military Leadership," *Army Information Digest* 8, no. 2 (February 1953): 5.

7-3 7-3 "The Army is people...": Lewis Sorley, *Thunderbolt: General Creighton Abrams and the Army of His Times* (New York: Simon and Schuster, 1992), 350.

7-8 "There is a soul...": Royle, 58.

7-8 7-42 "You have a...": H. B. Simpson, *Audie Murphy: American Soldier* (Dallas, TX: Alcor Publishing Co., 1982), 271.

7-8 7-43 "...give to the people...": Christopher J. Anderson, "Dick Winters: Reflections on the Band of Brothers, D-Day and Leadership," *American History Magazine* (August 2004): <http://www.historynet.com/magazines/american_history/3029766.html>.

7-9 7-45 "NSDQ": Mark Bowden, "Blackhawk Down," Chapter 29, *Philadelphia Inquirer* (14 December 1997): <http://inquirer.philly.com/packages/somalia/sitemap.asp>.

7-10 **One Man Can Make a Difference**: Eric Schmitt, "Medal of Honor to Be Awarded to Soldier Killed in Iraq, a First," *The New York Times* (30 March 2005): A13. Joe Katzman, "Medal of Honor: SFC Paul Ray Smith," Winds of Change Web site (23 October 2003): <http://www.windsofchange.net/archives/004196.php>.

7-10 "Leading and caring ...": *The Chiefs of Staff, United States Army: On Leadership and the Profession of Arms* (Washington, DC: Information Management Support Center, August, 2000), 37 (hereafter referred to as *Chiefs of Staff* 2000).

7-14 7-86 "The Commanding General...": Patton, 397–8.

Chapter 8: Developing

8-4 **SGT York**: David D. Lee, *Sergeant York: An American Hero* (Lexington, KY: The University Press of Kentucky, 1985), 33-38.

8-9 "...[G]ood NCOs are not...": William A. Connelly, "Keep Up with Change in the '80s," *Army Magazine* (October 1982): 29.

8-11 8-63 "The instruments of battle...": Colonel Charles Jean Jacques Ardant du Picq, *Battle Studies: Ancient and Modern Battle* (Carlisle Barracks, PA: U.S. Army War College, 1983), 68.

8-11 "Soldiers learn...": Richard A. Kidd, "NCOs Make It Happen," *Army Magazine* (October 1994): 34.

8-15 "The cohesion that...": *Chiefs of Staff* 2000, 6.

8-18 8-101 "The war brings ...": T. Skeyhill, ed., *Sergeant York: His Own Life Story and War Diary* (Garden City, NY: Doubleday, Doran. 1928), 212.

Chapter 9: Achieving

9-1 "It is in the minds...": Robert A Fitton, ed., *Leadership: Quotations from the Military Tradition* (Boulder, CO: Westview Press, 1990), 75.

9-4 "...[A] good plan...": Patton, 354.

9-5 "...[S]chools and their...": William G. Bainbridge, "Quality, Training and Motivation," *Army Magazine* (October 1976): 28.

9-7 "The American people...": Gregory Fontenot, E. J. Degen, and David Tohn, *On Point: The United States Army in Operation Iraqi Freedom* (Fort Leavenworth, KS: Combat Studies Institute Press), 5.

9-8 **Achieving Success and Leadership Excellence**: Jack J. Gifford, "Invoking Force of Will to Move the Force," *Studies in Battle Command* (Fort Leavenworth, KS: Combat Studies Institute, U.S. Army Command and General Staff College, 1995), 143-46.

Chapter 10: Influences on Leadership

10-1 "The role of leadership...": *Chiefs of Staff* 2000, 99.

10-2 10-14 Stefan Lovgren, "English in Decline as a First Language, Study Says," *National Geographic News* (26 February 2004): <http://news.nationalgeographic.com/news/2004/02/0226_040226_language.html>. Mike Bergman, "Nearly 1-in-5 Speak a Foreign Language at Home; Most Also Speak English 'Very Well,'" *U.S. Census Bureau News* (8 October 2003): <http://www.census.gov/Press-Release/www/releases/archives/census_2000/001406.html> .

10-3 10-17 Virtual Team. J.E. Driskell, P.H. Radtke, and E. Salas, "Virtual Teams: Effects of Technological Mediation on Team Performance," *Group Dynamics: Theory Research and Practice* (2003), 297-323.

10-4 "All men are...": Patton, 340.

10-4 10-29 Stress countermeasures. LTC Carl A. Castro and COL Charles W. Hoge, *10 Unpleasant Facts about Combat and What Leaders Can Do to Change Them* (Silver Spring, MD: Walter Reed Army Institute of Research, 31 August 1999).

10-5 "Sure I was ...": Donna Miles, "The Women of Just Cause," *Soldiers Magazine* (March 1990): 23.

10-6 **A Fearless Leader-Twice a Hero**: James B. Stewart, "The Real Heroes are Dead," *The New Yorker* (11 February 2002): <http://www.newyorker.com/fact/content?020211fa_FACT1>. Greyhawk, "911 Remembered: Rick Rescorla was a Soldier," The Mudville Gazette Web site (September 2003): <http://www.mudvillegazette.com>. Michael Grunwald, "A Tower of Courage," *The Washington Post* (28 October 2001): F01.

10-7 "War makes...": Royle, 55.

10-7 10-48 Tools for Adaptability: S. S. White, R. A. Mueller-Hanson, D. W. Dorsey, E. D. Pulakos, M. M. Wisecarver, E. A. Eagle, III, and K. G. Mendini, *Developing Adaptive Proficiency in Special Forces Officers* Research Report 1831 (Arlington, VA: U.S. Army Research Institute for the Behavioral and Social Sciences, 2005), 2.

10-8 10-53 "Nothing in the world...": Coffey, 248.

PART FOUR LEADING AT ORGANIZATIONAL AND STRTEGIC LEVELS

Chapter 11: Organizational Leadership

11-1 "The American Soldier...": Lt. General Lucian K. Truscott, Jr., *Command Missions: A Personal Story* (New York: E. P. Dutton and Company, Inc., 1954), 556.

11-3 "If you are the leader...": Gordon R. Sullivan and Michael V. Harper, *Hope is Not a Method: What Business Leaders Can Learn from America's Army* (New York: Crown Publishing Group, 1996), 232-233 (hereafter referred to as Sullivan).

11-3 "Too often we place...": *Chiefs of Staff* 1997, 7.

11-4 "It is not enough...": U.S. Army Command Information Unit, *Quotes for the Military Writer* (Washington, DC: Office of the Chief of Information, Department of the Army, 1972), 13-1. (hereafter known as *Military Quotes 1972).*

11-6 "When a team...": Joe Paterno in Lewis D. Eigen and Jonathan P. Siegel, *The Manager's Book of Quotations* (New York: The American Management Association, 1989), 471.

11-6 "Never tell...": Patton, 357.

11-10 **Joint and Combined Synchronization during Operation Assured Response**: Gil High, "Liberia Evacuation," *Soldiers Magazine* (July 1996): 4-5. John W. Partin and Rob Rhoden, *Operation Assured Response: SOCEUR's NEO in Liberia, April 1996* (Headquarters, US Special Operations Command, History and Research Office, Sep. 1997).

Chapter 12: Strategic Leadership

12-1 "Final decisions...": Peter G. Tsouras, ed., *The Greenhill Dictionary of Military Quotations* (Mechanicsburg, PA: Stackpole Books, 2000), 133.

12-2 "Leadership is...": Karel Thomas M. McNichols, Jr., Anthony J. Cotti, Jr., Thomas H. Hutchinson III, and Jackie Eckart Wehmueller, eds., *Naval Leadership: Voices of Experience* (Annapolis, MD: Naval Institute Press, 1987), 16.

12-2 "A tremendous amount...": Gordon R. Sullivan, "Strategic Change: The Way Forward," United States Department of Defense Web site (24 April 1995): <http://www.defenselink.mil/speeches/1995/s19950424-sullivan.html>.

12-3 **From Vision to Victory**: Richard M. Swain, *Lucky War: Third Army in Desert Storm* (Fort Leavenworth, KS: U.S. Army Command and General Staff College Press, 1997), 71-138. Carl H. Builder, Steven C. Bankes, and Richard Nordin, *Command Concepts: A Theory Derived from the Practice of Command and Control* (Santa Monica, CA: RAND Corporation, November 1999), 25-42. United States Central Command. *Operation Desert Shield/Desert Storm Executive Summary.* Unclassified Document (Tampa, FL: U.S. Central Command, 11 July 1991), 10-11. H. Norman Schwarzkopf, *It Doesn't Take a Hero* (New York: Bantam Books, 1992), 380.

12-3 12-14 "Once a vision...": Sullivan, 91.

12-3 **Combat Power from a Good Idea**: Judith A. Bellafaire, *The Women's Army Corps: A Commemoration of World War II Service* CMH Publication 72-15, (Washington, DC: Department of the Army, Center of Military History), <http://www.army.mil/cmh-pg/brochures/wac/wac.htm>.

Cynthia F. Brown, Major AN, Army Nurse Corps historian, SUBJECT: "Women in Leadership," memorandum, 4 November 1998.

12-6 "If you don't...": General Eric K. Shinseki, "Prepared Remarks General Eric K. Shinseki, Chief of Staff, United States Army, at the Association of the United States Army Seminar," U. S. Army news release (Washington, DC, 8 November 2001). From U. S. Army Web site, <http://www4.army.mil/ocpa/print.php?story_id_key=1417>.

12-7 "Difficulties mastered ...": R. Dale Jeffery. *The Soldier's Quote Book* (Houston, TX: DaVinci Publishing Group, 1999), 197.

12-9 "A good soldier...": *Military Quotes 1972*, 18-3. [Graduation Speech at the United States Military Academy, 17 June 1933, MacArthur Memorial Records Group 25 (Addresses, Statements and Speeches), Norfolk VA].

12-10 "The crucial military...": James Thomas Flexner, *George Washington in the American Revolution (1775-1783)* (Boston: Little, Brown, and Company, 1968), 535.

12-11 "From an intellectual...": William J. Crowe, Jr. with David Chanoff, *The Line of Fire: From Washington to the Gulf, the Politics and Battles of the New Military* (New York: Simon & Schuster, 1993), 54.

12-11 "Certainly one of the reasons...": Edgar F. Puryear, Jr., *Nineteen Stars: A Study in Military Character and Leadership* (Novato, CA: Presidio Publishing, 1994), 101.

12-13 "Continuity and change ...": *Chiefs of Staff* 2000, 24-25.

12-14 **The Quickest and Most Efficient Way to Plan**: Paul R. Howe, *Leadership and Training for the Fight* (New York: Skyhorse Publishing, 2011), 170-2.

12-16 "You will be informing...": Gordon R. Sullivan, General CSA, SUBJECT: "Reshaping Army Doctrine," memorandum for Lieutenant General Frederick M. Franks, Jr., 29 July 1991 quoted in Gregory Fontenot, E. J. Degen, and David Tohn, *On Point: The United States Army in Iraqi Freedom* (Fort Leavenworth, KS: Combat Studies Institute Press, 2003), 6.

Glossary

SECTION I – ACRONYMS AND ABBREVIATIONS

24/7	24 hours a day, 7 days a week
AAR	after-action review
AR	Army regulation
BG	brigadier general
COL	colonel
CPL	corporal
CPT	captain
DA	Department of the Army
DA Pam	Department of the Army Pamphlet
DOD	Department of Defense
FM	field manual
GEN	general
GPS	global positioning system
GTA	graphic training aid
HMMWV	high-mobility, multipurpose wheeled vehicle
IDP	individual development plan
IPR	in-process review
JP	joint publication
LDRSHIP	An aid for remembering the Army Values (loyalty, duty, respect, selfless service, honor, integrity, personal courage).
LT	lieutenant
LTC	lieutenant colonel
MAJ	major
NATO	North Atlantic Treaty Organization
NCO	noncommissioned officer
NCOER	noncommissioned officer evaluation report
OER	officer evaluation report
PVT	private
SFC	sergeant first class
SGT	sergeant
SHAEF	Supreme Headquarters Allied Expeditionary Force
STP	soldier training publication
SPC	specialist
SSG	staff sergeant
TOW	tube-launched, optically tracked, wire-guided (refers to a heavy antitank missile system)

U.S.	United States
USS	United States Ship
WAAC	Women's Army Auxiliary Corps
WO1	warrant officer 1

SECTION II – TERMS AND DEFINITIONS

***adaptability** An effective change in behavior in response to an altered situation.

***Army leader** Anyone who by virtue of assumed role or assigned responsibility inspires and influences people to accomplish organizational goals.Army leaders motivate people both inside and outside the chain of command to pursue actions, focus thinking, and shape decisions for the greater good of the organization.

Army Values Principles, standards, and qualities considered essential for successful Army leaders.

attribute Characteristic unique to an individual that moderates how well learning and performance occur.

climate The state of morale and level of satisfaction of members of an organization.

coaching The guidance of another person's development in new or existing skills during the practice of those skills.

command The authority that a commander in the military service lawfully exercises over subordinates by virtue of rank or assignment. Command includes the leadership, authority, responsibility, and accountability for effectively using available resources and planning the employment of, organizing, directing, coordinating, and controlling military forces to accomplish assigned missions. It includes responsibility for unit readiness, health, welfare, morale, and discipline of assigned personnel. (FMI 5-0.1)

commander's intent A clear, concise statement of what the force must do and the conditions the force must meet to succeed with respect to the enemy, terrain, and desired end state. (FM 3-0)

***core leader competencies** Groups of related leader behaviors that lead to successful performance, common throughout the organization and consistent with the organization's mission and values. What leaders should do to influence individual and organizational success.

***counseling** The process used by leaders to review with a subordinate the subordinate's demonstrated performance and potential.

critical thinking A deliberate process of thought whose purpose is to discern truth in situations where direct observation is insufficient, impossible or impractical.

culture The set of long-held values, beliefs, expectations, and practices shared by a group that signifies what is important and influences how an organization operates.

***direct leadership** The type of leadership that occurs at the smallest units of an organization and that is performed by leaders in first-line positions.

***domain knowledge** The body of facts, beliefs, and logical assumptions that people possess and use in areas of their work.

ethical reasoning	A type of reasoning that is characterized by beliefs of right and wrong and that applies in thinking and in the Army problem solving model. Three ethical perspectives are often combined in ethical reasoning: **virtues-based** – choices are based on desirable qualities like courage, justice, compassion. **principles-based** – choices are based on beliefs recognized by a group as authoritative or normative such as the seven Army Values, uniform code of military justice, or constitutional rights. **consequences-based** – choices are based on the action that produces the greatest good for the greatest number of people.
***informal leadership**	A type of leadership that is not based on command or other designation of formal authority. Informal leadership occurs as an individual exerts influence others for the good of the organization.
leader development	The deliberate, continuous, sequential, and progressive process, grounded in the Army Values, that grows Soldiers and civilians into competent and confident leaders capable of decisive action. (FM 7-0)
***leadership**	The process of influencing people by providing purpose, direction, and motivation, while operating to accomplish the mission and improve the organization.
***leader teams**	A group of leaders who are bound together by similar functions, tasks, organizational structure, or interests.
lifelong learning	The individual lifelong choice to actively and overtly pursue knowledge, the comprehension of ideas, and the expansion of depth in any area in order to progress beyond a known state of development and competency. (FM 7-0)
mental agility	A flexibility of mind, a tendency to anticipate or adapt to uncertain or changing situations.
mentorship	The voluntary developmental relationship that exists between a person of greater experience and a person of lesser experience that is characterized by mutual trust and respect. (AR 600-100)
military bearing	The projection of a commanding presence and a professional image of authority.
mission command	The conduct of military operations through decentralized execution based upon mission orders for effective mission accomplishment. Successful mission command results from subordinate leaders at all echelons exercising disciplined initiative within the commander's intent to accomplish missions. It requires an environment of trust and mutual understanding. (FM 6-0)
multisource assessment	A formal measure of peer, subordinate, superior, and self impressions of specified qualities of a single individual. Also called a multirater assessment, a 360 degree assessment or 360 appraisal (360 applies when all sources of ratings are collected).
officership	A particular type of leadership that is associated with the rank that a commissioned officer holds.
***organizational leadership**	The type of leadership that occurs at intermediate sized units of an organization such as brigade through corps levels or directorate through installation levels.

profession of arms	The vocation ascribed to all whose work involves mastery of the disciplined and open, collective application of force in pursuit of public purposes.
resilience	Tendency to recover quickly from setbacks, shock, injuries, adversity, and stress while maintaining a mission and organizational focus.
responsibility	(joint) The obligation to carry forward an assigned task to a successful conclusion. With responsibility goes authority to direct and take the necessary action to ensure success. (JP 1-02)
role	The functions and activities assigned to, required of or expected of a person or group.
self-awareness	Being aware of oneself, including one's traits, feelings, and behaviors.
self-development	A planned, continuous, life-long process individual leaders use to enhance their competencies and potential for progressively more complex and higher–level assignments. (DA Pam 350-58)
self-efficacy	A person's confidence in his or her ability to succeed at a task or reach a goal.
shared leadership	The sharing of authority and responsibility between two or more leaders for decision- making, planning, and executing.
***strategic leadership**	The type of leadership that occurs at the highest levels of the organization.
virtual team	Any team whose interactions are mediated by time, distance, or technology.
well-being	The personal, physical, material, mental, and spritual state of Soldiers, civilians, and their families that contributes to their preparadness to perform the Army's missions.

References

REQUIRED PUBLICATIONS

These documents must be available to intended users of this publication.
None

RELATED PUBLICATIONS

These sources contain relevant supplemental information.

JOINT PUBLICATIONS

Joint publications are available online: http://www.dtic.mil/doctrine/jel/

Joint Doctrine Capstone and Keystone Primer, 10 September 2001.

JP 0-2. *Unified Actions Armed Forces*, 10 July 2001.

JP 1. *Joint Warfare of the Armed Forces of the United States*, 14 November 2000.

JP 1-02. *Department of Defense Dictionary of Military and Associated Terms*, 12 April 2001.

JP 3-0. *Doctrine for Joint Operations*. 10 September 2001.

ARMY PUBLICATIONS

Army doctrinal publications are available online: http://www.apd.army.mil/

Army Regulations

AR 350-1. *Army Training and Leader Development*. 13 January 2006.

AR 600-20. *Army Command Policy*. 07June 2006.

AR 600-100. *Army Leadership*. 17 September 1993.

AR 601-280. *Army Retention Program*, 31 January 2006.

AR 623-3. *Evaluation Reporting System*. 15 May 2006.

AR 690-11. *Use and Management of Civilian Personnel in Support of Military Contingency Operations*. 26 May 2004.

Department of the Army Pamphlets and Graphic Training Aids

DA Pam 350-58. *Leader Development for America's Army*. 13 October 1994.

DA Pam 600-3. *Commissioned Officer Professional Development and Career Management*. 28 December 2005.

DA Pam 600-25. *U.S. Army Noncommissioned Officer Professional Development Guide*. 15 October 2002.

DA Pam 600-69. *Unit Climate Profile Commander's Handbook*. 1 October 1986.

GTA 22-6-1. Ethical Climate Assessment Survey.

Field Manuals

FM 1. *The Army*. 14 June 2005.

FM 3-0. *Operations*. 14 June 2001.

FM 3-07. *Stability Operations and Support Operations*. 20 February 2003.

FM 3-07.31. *Peace Operations Multi-service Tactics, Techniques, and Procedures for Conducting Peace Operations*. 26 October 2003.

FM 3-13 (100-6). *Information Operations: Doctrine, Tactics, Techniques, and Procedures*. 28 November 2003.

FM 3.50-1. *Army Personnel Recovery.* 10 Aug 2005.

FM 3-90. *Tactics.* 4 July 2001.

FM 3-100.12. *Risk Management: Multiservice Tactics, Techniques, and Procedures For Risk Management.* 15 February 2001.

FM 3-100.21. *Contractors on the Battlefield.* 3 January 2003.

FM 3-100.38. *UXO Multi-service Tactics, Techniques, and Procedures for Unexploded Ordnance Operations.* 18 August 2005.

FM 5-0 (101-5). *Army Planning and Orders Production.* 20 January 2005.

FM 6-0. *Mission Command: Command and Control of Army Forces.* 11 August 2003.

FM 6-22.5. *Combat Stress.* 23 June 2000.

FM 7-0 (25-100). *Training the Force.* 22 October 2002.

FM 7-1 (25-101). *Battle Focused Training.* 15 September 2003.

FM 7-22.7. *The Army Noncommissioned Officer Guide*, 23 December 2002.

FM 4-02.51. *Combat and Operational Stress Control.* 6 July 2006.

FM 21-20. *Physical Fitness Training.* 30 September 1992.

FM 71-100. *Division Operations*, 28 August 1996. FM 71-100 will be republished as FM 3-91.

FM 100-7. *Decisive Force: The Army in Theater Operations.* 31 May 1995. FM 100-7 will be republished as FM 3-93.

FM 100-15. *Corps Operations*, 29 October 1996. FM 100-15 will be republished as FM 3-92.

FMI 3-63.6. *Command and Control of Detainee Operations.* 23 September 2005.

DOD Civilian Leader Policies

DOD directives are available online at http://www.dtic.mil/whs/directives

DOD 1400.25-M. *DOD Civilian Personnel Manual.* December 1996. Changes 1–19.

DODD 1400.5. *DOD Policy for Civilian Personnel.* 12 January 2005.

DODD 1400.6. *DOD Civilian Employees in Overseas Areas.* 15 February 1980.

DODD 1430.2. *Civilian Career Management,* 13 June 1981. Change 1, 16 November 1994.

DODD 1430.4. *Civilian Employee Training*, 30 January 1985. Change 1, 16 November 1994.

DODD 1430.14. *DOD Executive Leadership Development Program (ELDP).* 12 September 2003.

DODI 3020.41. *Contractor Personnel Authorized to Accompany the U.S. Armed Forces.* 3 October 2005.

Soldier Training Publications

STP 21-1-SMCT. *Soldier's Manual of Common Tasks Skill Level 1.* 2 October 2006.

TRADOC Pamphlets

TRADOC Pam 525-100-4. *Leadership and Command on the Battlefield: Noncommissioned Officer Corps.* 1994.

NONMILITARY PUBLICATIONS

"African-American Vet Receives Medal of Honor." Home of Heroes Web site, 13 January 1997. <http://www.homeofheroes.com/news/archives> scroll to 1997, select Jan 13.

Anderson, Christopher J. "Dick Winters' Reflections on His Band of Brothers, D-Day and Leadership." *American History Magazine*, August 2004. <http://www.historynet.com/magazines/american_history/3029766.html>.

Ardant du Picq, Charles Jean Jacques Joseph. *Battle Studies: Ancient and Modern Battle*. Translated by John W. Greely and Robert C. Cotton. Harrisburg, PA: Military Service Publishing Co., 1947.

Aristotle. *Nicomachean Ethics*. Translated by Martin Ostwald. New York: Macmillan Publishing Company, 1962.

Arlington National Cemetery. "Remains Never Recovered." Arlington National Cemetery Web site. <http://www.arlingtoncemetery.net/medalofh.htm>, scroll to Humbert Roque Versace, USA.

"Army Awards MPs for Turning Table on Ambush." *Army News Service*, 16 June 2005. <http://www4.army.mil/ocpa/read.php?story_id_key=7472>.

Bainbridge, William G. "Quality, Training and Motivation." *Army Magazine*, October 1976.

Bandura, Albert. "Self-efficacy." *Encyclopedia of Human Behavior* (Vol 4). San Diego, CA: Academic Press, 1994.

Barnes, Julian E. "A Thunder Run Up Main Street." *U.S. News and World Report*, 14 April 2003: <http://www.usnews.com/usnews/news/articles/030414/14front_2.htm>.

Bartlett, John, ed. *Familiar Quotations: A Collection of Passages, Phrases and Proverbs Traced to Their Sources in Ancient and Modern Literature*. Boston: Little, Brown and Company, 1968.

Bellafaire, Judith L. *The Women's Army Corps: A Commemoration of World War II Service*. CMH Publication 72-15. Washington, DC: Department of the Army, Center of Military History. <http://www.army.mil/cmh-pg/brochures/wac/wac.htm>.

Bergman, Mike. "Nearly 1-in-5 Speak a Foreign Language at Home; Most Also Speak English Very Well." *U.S. Census Bureau News,* 8 October 2003. <http://www.census.gov/Press-Release/www/releases/archives/census_2000/001406.html>.

Black, Edwin. *Banking on Baghdad: Inside Iraq's 7,000-Year History of War, Profit, and Conflict*. New York: John Wiley and Sons, 2004.

Blumenson, Martin. "Task Force Kingston." *Army Magazine*, April 1964.

Bowden, Mark. "Blackhawk Down," Chapter 29. *Philadelphia Inquirer*, 14 December 1997.

Bradley, Omar N. "American Military Leadership." *Army Information Digest* 8, no. 2, February 1953.

_____. "Leadership: An Address to the U.S. Army War College, 07 Oct 1971." *Parameters* 1 (3), 1972.

Briscoe, Jon P., and Douglas T. Hall. "Grooming and Picking Leaders Using Competency Frameworks: Do They Work? An Alternative Approach and New Guidelines for Practice." *Organizational Dynamics*, 28 (1999): 37-52.

Brown, Cynthia F, Major AN, Army Nurse Corps Historian., SUBJECT: "Women in Leadership." Memorandum, 4 November 1998.

Brown, Frederic J. *Vertical Command Teams*. IDA Document D-2728. Alexandria, VA: Institute for Defense Analyses, 2002.

Builder, Carl, Steven C. Bankes, and Richard Nordin. *Command Concepts: A Theory Derived from the Practice of Command and Control*. Santa Monica, CA: RAND Corporation, November 1999.

Burk, Robin. "Tell Me Again About Women in Combat." Winds of Change Web site, 25 March 2005. <http://www.windsofchange.net/archives/006564.php>.

Busetti, Linda. "Local Vietnam War Hero Receives Medal of Honor." *Arlington Catholic Herald*, 11 July 2002.

Butcher, Harry C. *My Three Years with Eisenhower: The Personal Diary of Captain Harry C. Butcher, USNR, Naval Aide to General Eisenhower, 1942 to 1945*. New York: Simon & Schuster, 1946.

"Capt. Humbert Roque "Rocky" Versace, Captured by Viet Cong in 1963 and Executed in 1965." Special Operations Memorial Web site, <http://www.somf.org/moh/>, scroll to Versace, H.R. "Rocky" USA.

Castro, LTC Carl A., and COL Charles W. Hoge. Briefing: *10 Unpleasant Facts about Combat and What Leaders Can Do to Change Them.* Silver Spring, MD: Walter Reed Army Institute of Research, 31 August 1999.

The Chiefs of Staff, United States Army: On Leadership and the Profession of Arms. Washington, DC: The Information Management Support Center, 1997.

The Chiefs of Staff, United States Army: On Leadership and the Profession of Arms. Washington, DC: The Information Management Support Center, 2000.

Coffey, William T., Jr., *Patriot Hearts: An Anthology of American Patriotism.* Colorado Springs, CO: Purple Mountain Publishing, 2000.

Combat Studies Institute. *Studies in Battle Command.* Fort Leavenworth, KS: U.S. Army Command and General Staff College, 1995.

Connelly, William. "NCO's: It's Time to Get Tough." *Army Magazine*, October 1981.

_____. "Keep up with Change in the 80's." *Army Magazine*, October 1982.

Crowe, Jr., William J. with David Chanoff. *The Line of Fire: From Washington to the Gulf, the Politics and Battles of the New Military.* New York: Simon & Schuster, 1993.

Davis, Burke. *They Called Him Stonewall.* New York: Rinehart & Company, Inc., 1954.

Davis, William C. *Battle at Bull Run: A History of the First Major Campaign of the Civil War.* Baton Rouge, LA: Louisiana State University Press, 1977.

DeWeerd, H.A., ed. *Selected Speeches and Statements of General of the Army George C. Marshall.* Washington, DC: The Infantry Journal, 1945.

Driskell, J.E., P.H. Radtke, and E. Salas, "Virtual Teams: Effects of Technological Mediation on Team Performance." *Group Dynamics: Theory Research and Practice* 7 (4) (2003): 297-323.

Eigen, Lewis D. *The Manager's Book of Quotations.* New York: The American Management Association, 1989.

Essens, Peter, Ad Vogelaar, Jacques Mylle, Carol Blendell, Carol Paris, Stan M. Halpin, and Joe Baranski. *Military Command Team Effectiveness: Model and Instrument for Assessment and Improvement.* NATO RTO Technical Report AC/323(HFM-087) TP/59. Soesterberg, Netherlands: TNO Human Factors, 2005.

Fallesen, Jon J., and Rebecca J. Reichard. *Leadership Competencies: Building a Foundation for Army Leader Development.* Los Angeles, CA: Society of Industrial and Organizational Psychologists, 2005.

Fitton, Robert A., ed. *Leadership: Quotations from the Military Tradition.* Boulder, CO: Westview Press, 1990.

Flexner, James Thomas. *George Washington in the American Revolution (1775-1783).* Boston: Little Brown and Company, 1968.

Fontenot, Gregory, E. J. Degen, and David Tohn. *On Point: The United States Army in Operation Iraqi Freedom.* Fort Leavenworth, KS: Combat Studies Institute Press, 2003.

Gates, Julius W. "From the Top." *Army Trainer* 9, no. 1, Fall 1989.

Gifford, Jack J. "Invoking Force of Will to Move the Force." *Studies in Battle Command.* Fort Leavenworth, KS: Combat Studies Institute, U.S. Army Command and General Staff College, 1995.

Grunwald, Michael. "A Tower of Courage." *The Washington Post*, 28 October 2001.

Gunlicks, James B. Acting Director of Training. SUBJECT: "Army Training and Leader Development Panel–Civilian (ATLDP-CIV), Implementation Process Action Team (IPAT) Implementation Plan–ACTION MEMORANDUM." Memorandum for Chief of Staff, Army, 28 May 2003.

Hannah, Dogen. "In Iraq, U.S. Military Women Aren't Strangers to Combat." *Knight Ridder Newspapers*, 7 April 2005.

Hall, Douglas T. "Self-Awareness, Identity, and Leader Development," David V. Day, Stephen J. Zaccaro, and Stanley M. Halpin, eds. *Leader Development for Transforming Organizations: Growing Leaders for Tomorrow*. Mahwah, NJ: Erlbaum, 2004.

Heenan, David A., and Warren Bennis. *Co-leaders: The Power of Great Partnerships*. New York: Wiley, 1999.

Heinl, Robert Debs, Jr. *Dictionary of Military and Naval Quotations*. Annapolis, MD: U.S. Naval Institute Press, 1966.

Hesselbein, Francis, ed. *Leader to Leader*. New York: Leader to Leader Institute, 2005.

High, Gil. "Liberia Evacuation." *Soldiers Magazine*, July 1996.

Horey, Jeffrey, Jon J. Fallesen, Ray Morath, Brian Cronin, Robert Cassella, Will Franks, Jr., and Jason Smith. *Competency Based Future Leadership Requirements* (Technical Report 1148). Arlington, VA: Army Research Institute for the Behavioral and Social Sciences, July 2004.

Horey, Jeffrey, Jennifer Harvey, Patrick Curtin, Heidi Keller-Glaze, and Jon J. Fallesen. *A Criterion-Related Validation Study of the Army Core Leader Competency Model*. (Technical Report). Arlington, VA: U.S. Army Research Institute for the Behavioral and Social Sciences, 2006.

Jeffery, R. Dale. *The Soldier's Quote Book*. Houston, TX: DaVinci Publishing Group, 1999.

Joseph and Edna Josephson Institute of Ethics. *Making Ethical Decisions*. Joseph and Edna Josephson Institute of Ethics Web site. <http://www.josephsoninstitute.org/MED/MED-intro+toc.htm>.

Katzman, Joe. "Medal of Honor: SFC Paul Ray Smith." Winds of Change Web site, 23 October 2003. <http://www.windsofchange.net/archives/004196.php>. Kelly, S. H. "Seven WWII Vets to Receive Medals of Honor." *Army News Service*, 13 January 1997. <http://www4.army.mil/ocpa/print.php?story_id_key=2187>.

Kem, Jack D. "A Pragmatic Ethical Decision Making Model for the Army: The Ethical Triangle" Text for L100: Leadership, Intermediate Level Education Common Core. Fort Leavenworth, KS: U.S. Army Command and General Staff College, 2005.

Kerr, Bob. "CGSC Class of 2005 Graduates." *Fort Leavenworth Lamp*, 23 June 2005.

Kidd, Richard A. "NCOs Make It Happen." *Army Magazine*, October 1994.

Kluever, Emil K., William L. Lynch, Michael T. Matthies, Thomas L. Owens, and John A. Spears. *Striking a Balance in Leader Development: A Case for Conceptual Competence*. National Security Program Discussion Paper Series, 92-02. Cambridge, MA: Harvard University, 1992.

Lacey, Jim. "From the Battlefield." *Time Magazine*, 14 April 2003: <http://www.time.com/time/magazine/article/0,9171,1004638,00.html>.

Lee, David D. *Sergeant York: An American Hero*. Lexington, KY: The University Press of Kentucky, 1985.

Locke, Edwin A. "The Motivation to Work: What We Know." *Advances in Motivation and Achievement,* 10 (1997): 375-412.

Lovgren, Stefan. "English in Decline as a First Language, Study Says." *National Geographic News*, 26 February 2004. <http://news.nationalgeographic.com/news/2004/02/0226_040226_language.html>.

Marshall, George C. *Quotes for the Military Writer*. Washington, DC: Department of the Army, Office of the Chief of Information, August 1972.

Marshall, S. L. A. *Men Against Fire: The Problem of Battle Command in Future War*. Gloucester, MA: Peter Smith, 1978.

Martin, James, ed. *The Military Quotation Book*. New York: St. Martin's Press, 1990.

Maxwell, John C. *Leadership 101–Inspirational Quotes and Insights for Leaders*. Tulsa, OK: Honor Books, 1994.

"Medic on a Mission: An Army Medics Strong-Arm Tactics Help to Carry the Day." Medal of Honor Web site, <http://www.medalofhonor.com/DavidBleak.htm>.

Miles, Donna. "The Women of Just Cause." *Soldiers Magazine*, March 1990.

Mitchell, T. R., and D. Daniels. "Motivation," W. C. Borman, D. R. Ilgen, and R. Klimoski, eds. *Handbook of Psychology* (Vol. 12: Industrial and Organizational Psychology). Hoboken, NJ: Wiley, 2003.

"MOH Citation for Humbert Roque Versace." Home of Heroes Web site, <http://www.homeofheroes.com/moh>, scroll to bottom of web page, click on Search Our Site, type in Versace in Google search, select MOH Citation for Humbert Roque Versace.

Montor, Karel, Thomas M. McNichols, Jr., Anthony J. Cotti, Jr., Thomas H. Hutchinson III, and Jackie Eckart Wehmueller, eds. *Naval Leadership: Voices of Experiences*. Annapolis, MD: U.S. Naval Institute Press, 1987.

Moss, James. *Noncommissioned Officers' Manual*. Menasha, WI: George Banta Publishing Co., 1917.

Olson, James S., and Randy Roberts. *My Lai: A Brief History with Documents*. Boston: Bedford Books, 1998.

"The Only Living African American World War II Hero to Receive the Medal of Honor." Medal of Honor Web site, 13 January 1997. <http://www.medalofhonor.com/VernonBaker.htm>.

The Noncom's Guide: An Encyclopedia of Information for All Noncommissioned Officers of the United States Army 16th ed. Harrisburg, PA: The Stackpole Company, 1962.

Partin, John W., and Rob Rhoden. *Operation Assured Response: SOCEUR's NEO in Liberia, April 1996*. Headquarters, U.S. Special Operations Command, History and Research Office, 1997.

Patton, George S., Jr. *War as I Knew It*. Boston: Houghton Mifflin Company, 1947.

Pearce, Craig L., and Henry P. Sims, Jr. "Vertical Versus Shared Leadership as Predictors of the Effectiveness of Change Management Teams: An Examination of Aversive, Directive, Transactional, Transformational, and Empowering Leader Behaviors." *Group Dynamics: Theory, Research, and Practice*, 6 (2002): 172-197.

Peters, W. R. *The My Lai Inquiry*. New York: W.W. Norton, 1979.

Pike, John. "Operation Enduring Freedom-Afghanistan." Global Security Web site, 7 March 2005. <http://www.globalsecurity.org/military/ops/enduring-freedom.htm>.

Powell, Colin L. *My American Journey*. New York: Random House, 1995.

Pullen, John J. *The Twentieth Maine*. Philadelphia: J.B. Lippincott Co., 1957. Reprint, Dayton, Ohio: Press of Morningside Bookshop, 1980.

Puryear, Edgar F., Jr. *Nineteen Stars: A Study in Military Character and Leadership*. Novato, CA: Presidio Publishing, 1994.

Royle, Trevor, ed. *Dictionary of Military Quotations*. New York: Simon & Schuster, 1990.

Schmitt, Eric. "Medal of Honor to be Awarded to Soldier in Iraq, a First," *The New York Times*, 30 March 2005.

Schofield, John M. *Manual for Noncommissioned Officers and Privates of Infantry of the Army of the United States*. West Point, NY: U.S. Military Academy Library Special Collections, 1917.

Schwarzkopf, H. Norman. *It Doesn't Take a Hero*. New York: Bantam Books, 1992.

Senge, Peter M. *The Fifth Discipline: The Art and Practice of the Learning Organization*. New York: Doubleday Currency, 1990.

"Seven Black Soldiers from WWII Tapped to Receive Medal of Honor." *Boston Sunday Globe*, 28 April 1996. <http://www.366th.org/960428.htm>.

Shinseki, Eric K., General, United States Army. SUBJECT: "Implementing Warrior Ethos for The Army." Memorandum for Vice Chief of Staff, Army, 3 June 2005.

_____. "Prepared Remarks General Eric K. Shinseki, Chief of Staff, United States Army, at the Association of the United States Army Seminar." U. S. Army news release. Washington, DC, 8 November 2001. From U. S. Army Web site, <http://www4.army.mil/ocpa/print.php?story_id_key=1417>.

Simpson, H. B. *Audie Murphy: American Soldier*. Dallas, TX: Alcor Publishing Co., 1982.

Skeyhill, T., ed. *Sergeant York: His Own Life Story and War Diary*. Garden City, NY: Doubleday, Doran, 1928.

Smith, Margaret Chase. Speech to graduating women naval officers at Naval Station. Newport, RI. Skowhegan, ME: Margaret Chase Smith Library, 1952.

Sorley, Lewis. *Thunderbolt: General Creighton Abrams and the Army of His Times*. New York: Simon and Schuster, 1992.

Spector, Ronald H. *Eagle Against the Sun*. New York: Random House, 1985.

Steele, Dennis. "Broadening the Picture Calls for Turning Leadership Styles." *Army Magazine* 39 no.12, December 1989.

Steele, William M., and Robert P. Walters, Jr. "21st Century Leadership Competencies." *Army Magazine,* Aug 2001.

Stewart, James B. "The Real Heroes are Dead." *The New Yorker*, 11 February 2005. <http://www.newyorker.com/fact/content?020211fa_FACT1>.

Sullivan, Gordon R., General CSA. SUBJECT: "Reshaping Army Doctrine." Memorandum for Lieutenant General Frederick M. Franks, Jr., 29 July 1991. Quoted in Gregory Fontenot, E. J. Degen, and David Tohn. *On Point: The United States Army in Iraqi Freedom.* Fort Leavenworth, KS: Combat Studies Institute Press, 2003.

_____. "Strategic Change: The Way Forward." United States Department of Defense Web site, 24 April 1995. <http://www.defenselink.mil/speeches/1995/s19950424-sullivan.html>.

Sullivan, Gordon R. and Michael V. Harper. *Hope is Not a Method: What Business Leaders Can Learn from America's Army*. New York: Crown Publishing Group, 1996.

Swain, Richard. M. *Lucky War: Third Army in Desert Storm.* Fort Leavenworth, KS: U.S. Army Command and General Staff College Press, 1997.

Taylor, A. J. P. and S. L. Mayer. *History of World War II.* London: Octopus Books, 1974.

Thompson, Mark. "Randal Perez Didn't Join the Army to be a Hero." *Time Magazine*, 1 September 2002. <http://www.time.com/time/covers/1101020909/aperez.html>.

Title 10 USC 3583. *Requirement of Exemplary Service and the Army Values*.

Truscott, Lt. General Lucian K. *Command Missions: A Personal Story*. New York: E. P. Dutton & Company, Inc., 1954.

Tsouras, Peter E., ed. *The Greenhill Dictionary of Military Quotations*. Mechanicsburg, PA: Stackpole Books, 2000.

Tyson, Ann Scott. "Anaconda: A War Story." *Christian Science Monitor*, 1 August 2002. <http://www.csmonitor.com/2002/0801/p01s03-wosc.htm>.

_____. "Soldier Earns Silver Star for Her Role in Defeating Ambush." *Washington Post,* 17 June 2005.

United States Central Command. *Operation Desert Shield/Desert Storm Executive Summary.* Unclassified Document. Tampa, FL: U.S. Central Command, 11 July 1991.

U. S. Department of Defense. "Interview with U.S. Army Soldiers who Participated in Operation Anaconda." United States Department of Defense Web site, 7 March 2002. <http://www.defenselink.mil/Transcripts/Transcript.aspx?TranscriptID=2914>.

Webb, James H. *A Country Such as This*. Annapolis, MD: Naval Institute Press, March 2001.

White, Susan S., Rose A. Mueller-Hanson, David W. Dorsey, Elaine D. Pulakos, Michelle M. Wisecarver, Edwin A. Eagle, III, and Kip G. Mendini. *Developing Adaptive Proficiency in Special Forces Officers*. Research Report 1831. Arlington, VA: U.S. Army Research Institute for the Behavioral and Social Sciences, 2005.

Wood, Sara Ann. "Female Soldier Receives Silver Star in Iraq." *American Forces Press Service*, 17 June 2005. <http://www4.army.mil/ocpa/print.php?story_id_key=7474>.

Woodward, Bob. *The Commanders*. New York: Pocket Star Books, 1991.

Yukl, Gary, Carolyn Chavez, and Charles F. Seifert. "Assessing the Construct Validity and Utility of Two New Influence Tactics." *Journal of Organizational Behavior*, 26 (2005): 1-21.

Yukl, Gary, and J. Bruce Tracey. "Consequences of Influence Tactics Used With Subordinates, Peers, and the Boss." *Journal of Applied Psychology*, 77, (1992): 525-535.

PRESCRIBED FORMS

DA Form 4856. *Developmental Counseling Form.*

REFERENCED FORMS

DA Form 67-9. *Officer Evaluation Report.*

DA Form 67-9-1. *Office Evaluation Report Support Form.*

DA Form 67-9-1A. *Developmental Support Form.*

DA Form 2028. *Recommended Changes to Publications and Blank Forms.*

SUGGESTED READINGS FOR ARMY LEADERS

The Professional Reading List is a way for leaders at all levels to increase their understanding of our Army's history, the global strategic context, and the enduring lessons of war. The topics and time periods included in the books on this list are expansive and are intended to broaden each leader's knowledge and confidence. I challenge all leaders to make a focused, personal commitment to read, reflect, and learn about our profession and our world. Through the exercise of our minds, our Army will grow stronger.

General Peter J. Schoomaker, Chief of Staff, U.S. Army

The following book lists represent selections recommended for leaders at direct, organizational, and strategic levels of leadership, conceptual foundations for leadership, and cultural and regional studies. The current U.S. Army Chief of Staff recommendations are marked with an asterisk (*). The CSA list can be found at http://www.us.army.mil/cmh-pg/reference/CSAList.

FOR DIRECT LEADERS

*The Constitution of the United States. Available at <http://www.house.gov/Constitution/Constitution.html.

*Ambrose, Stephen E. *Band of Brothers: E Company, 506th Regiment, 101st Airborne from Normandy to Hitler's Eagle's Nest*. New York: Touchstone, 2001.

Applegate, Rex. *Kill or Get Killed*. Boulder, CO: Paladin Press, 2002.

*Appleman, Roy E. *East of Chosin: Entrapment and Breakout in Korea, 1950*. College Station, TX: Texas A&M Press, 1991.

*Atkinson, Rick. *An Army at Dawn: The War in Africa, 1942–1943*. New York: Henry Holt and Company, LLC, 2002.

Battle of Algiers. Movie. Video: Rome, Italy: Igor Films, 1966. DVD: Irvington, NY: Criterion Collection, 2004.

*Bergerud, Eric M. *Touched with Fire: The Land War in the South Pacific*. New York: Penguin Books, 1996.

*Berkowitz, Bruce. *The New Face of War: How War Will Be Fought in the 21st Century*. New York: The Free Press, 2003.

Burns, James MacGregor. *Leadership*. New York: Harper & Row, 1978.

Chamberlain, Joshua Lawrence. *The Passing of the Armies*. Dayton, OH: Press of Morningside Bookshop, 1981.

Clarke, Bruce C. *Guidelines for the Leader and the Commander*. Harrisburg, PA: Stackpole Books, 1973.

Clausewitz, Carl von. *On War*. Edited and translated by Michael Howard and Peter Paret. Princeton, NJ: Princeton University Press, 1976.

*Coffman, Edward M. *The War to End All Wars: The American Military Experience in World War I*. Lexington, KY: The University of Kentucky Press, 1986.

*D'Este, Carlos. *Patton: A Genius for War*. New York: HarperCollins Publishers, 1995.

*Doubler, Michael D. *Closing with the Enemy: How GIs Fought the War in Europe, 1944–1945*. Lawrence, KS: University of Kansas Press, 1994.

*Durant, Michael J. with Steven Hartov. *In the Company of Heroes*. New York: New American Library, 2003.

Fisher, Ernest F., Jr. *Guardians of the Republic: A History of the Non-Commissioned Officer Corps of the U.S. Army*. New York: Ballantine Books; 1994.

Forester, C. S. *Rifleman Dodd*. Garden City, NY: Sun Dial Press, 1944.

Gabriel, Richard A. *To Serve with Honor: A Treatise on Military Ethics and the Way of the Soldier*. Westport, CT: Greenwood Press, 1982.

Galula, David. *Counterinsurgency Warfare: Theory and Practice*. New York: Praeger, 1964. Hailer Publishing Paperback reprint, 2005.

*Grant, Ulysses S. *Personal Memoir: Ulysses S. Grant*. New York: Random House, 1999. Reprint of *Personal Memoirs of U.S. Grant*. New York: Charles Webster & Company, 1885.

Grossmann, Dave. *On Killing*. New York: Back Bay Books, 1996.

_____. *On Combat*. Portland, OR: PPCT Research Publications, 2004.

*Heller, Charles E., and William A. Stofft. *America's First Battles: 1776–1965*. Lawrence, KS: University Press of Kansas, 1986.

*Hogan, David W. *Centuries of Service: The U.S. Army 1775-2004*. (CMH Pub. 70–71–1). Washington, DC: Center of Military History, 2004. <http://www.army.mil/cmh-pg/books/COS/index.htm>.

Holmes, Richard. *Acts of War: The Behavior of Men in Battle*. New York: Free Press, 1985.

Jacobs, Bruce. *Heroes of the Army: The Medal of Honor and its Winners*. New York: W.W. Norton & Co., 1956.

*Keegan, John. *The Face of Battle: A Study of Agincourt, Waterloo, and the Somme*. New York: Viking Press, 1976. Reprint, New York: Penguin Books, 1978.

Kellett, Anthony. *Combat Motivation: The Behavior of Soldiers in Battle*. Boston: Kluwer-Nijhoff Publishing, 1982.

*Kindsvatter, Peter S. *American Soldiers: Ground Combat in the World Wars, Korea, and Vietnam*. Lawrence, KS: University Press of Kansas, 2003.

Kipling, Rudyard. *The Man Who Would Be King, and Other Stories*. Oxford: Oxford University Press, 1999. Video: Burbank, CA: Warner Home Video, 1975.

*Kolenda, Christopher, ed. *Leadership: The Warrior's Art*. Carlisle, PA: The Army War College Foundation Press, 2001.

*Linn, Brian McAllister. *The Philippine War, 1899–1902*. Lawrence, KS: University Press of Kansas, 2002.

MacDonald, Charles B. *The Battle of the Huertgen Forest*. New York: J. P. Lippincott Co., 1963.

*_____. *Company Commander*. New York: Bantam Books, 1979.

Malone, Dandridge M. *Small Unit Leadership*. Novato, CA: Presidio Press, 1983.

Matthews, Lloyd J. *The Challenge of Military Leadership*. New York: Pergamon-Brassey's International Defense Publishers, Inc., 1989.

*Millett, Allan R., and Peter Maslowski. *For the Common Defense: A Military History of the United States of America.* New York: The Free Press, 1994.

*Moore, Harold G., and Joseph L. Galloway. *We Were Soldiers Once...and Young.* New York: Random House, 1992.

Morgan, Forrest E. *Living the Martial Way: A Manual for the Way a Modern Warrior Should Think.* Fort Lee, NJ: Barricade Books, 1992.

Myrer, Anton. *Once an Eagle.* New York: Dell Publishing Co., 1970.

Naylor, Sean. *Not a Good Day to Die.* New York: The Berkley Publishing Group, 2005.

Newman, Aubrey S. *Follow Me.* San Francisco: Presidio Press, 1981.

Norton, Oliver Willcox. *The Attack and Defense of Little Round Top.* Dayton, OH: Press of Morningside Bookshop, 1978.

*Nye, Roger H. *The Challenge of Command: Reading for Military Excellence.* New York: The Berkley Publishing Group, 1986.

Pressfield, Steven. *Gates of Fire.* New York: Bantam Books, 1998.

Pullen, John J. *The Twentieth Maine.* Philadelphia: J.B. Lippincott Co., 1957. Reprint, Dayton, OH: Press of Morningside Bookshop, 1980.

Sajer, Guy. *The Forgotten Soldier.* New York: Harper & Row, 1971.

Shaara, Michael. *The Killer Angels.* New York: Ballantine Books, 1975.

Smith, Perry M. *Taking Charge: A Practical Guide for Leaders.* Washington, DC: National Defense University Press, 1986.

Small Wars Manual 1940 FMFRP 12-15. Marine Corps Command. Quantico, VA: U.S. Government Printing Office, 1940. <http://www.smallwars.quantico.usmc.mil/sw_manual.asp>.

Small Wars / 21st Century Addendum 2005. Quantico, VA: Marine Corps Combat Development Command, 2005.

Stockdale, James B. *A Vietnam Experience: Ten Years of Reflection.* Stanford, CA: Hoover Press, 1984.

Von Schell, Adolf. *Battle Leadership.* Columbus, GA: The Benning Herald, 1933.

Webb, James. *Fields of Fire.* New York: Bantam Books, 1985.

*Wilson, George D. *If You Survive: From Normandy to the Battle of the Bulge to the End of World War II.* New York: Ballantine Books, 1987.

FOR ORGANIZATIONAL AND STRATEGIC LEADERS

*National Security Strategy of the United States of America. <http://www.whitehouse.gov/nsc/nss.pdf>.

*National Strategy for Combating Terrorism. <http://www.whitehouse.gov/news/releases/2003/02/counter_terrorism/counter_terrorism_strategy.pdf>.

Ardant du Picq, Charles Jean Jacques Joseph. *Battle Studies: Ancient and Modern.* Translated by John W. Greely and Robert C. Cotton. Harrisburg, PA: Military Service Publishing Co., 1947.

*Bennis, Warren. *On Becoming a Leader.* Cambridge, MA: Perseus Publishing, 2003.

Blair, Clay. *The Forgotten War.* New York: Doubleday, 1987.

Boot, Max. *Savage Wars of Peace: Small Wars and the Rise of American Power.* New York: Basic Books, 2002.

Cecil, Hugh, and Peter Liddle. *Facing Armageddon: The First World War Experience.* South Yorkshire, UK: Pen & Sword Books, 2003.

Chilcoat, Richard A. *Strategic Art: The New Discipline for 21st Century Leaders.* Carlisle Barracks, PA: U.S. Army War College, Strategic Studies Institute, 1995.

Clancy, Tom. *Into the Storm.* New York: G. P. Putnam's Sons, 1997.

*Clausewitz, Carl von. *On War*. Edited and translated by Michael Howard and Peter Paret. New York: Alfred A. Knopf, 1976.

Davis, Burke. *The Campaign that Won America: The Story of Yorktown*. New York: The Dial Press, 1970.

Fehrenbach, R. R. *This Kind of War: A Study in Unpreparedness*. New York: Macmillan Co., 1963.

Freeman, Douglas Southall. *Lee's Lieutenants: A Study in Command*. 3 vols. New York: Charles Scribner's Sons, 1942-44.

*Friedman, Thomas. *The Lexus and the Olive Tree: Understanding Globalization*. New York: Anchor Books, 2000.

_____. *The World Is Flat: A Brief History of the Twenty-first Century*. New York: Farrar, Straus and Giroux, 2005.

Freytag-Loringhoven, Hugo F .P. J. von. *The Power of Personality in War*. In Art of War Colloquium text. Carlisle Barracks, PA: U.S. Army War College, September 1983.

Fuller, J. F. C. *The Conduct of War 1789-1961*. New Brunswick, NJ: Rutgers University Press, 1961.

Gaddis, John Lewis. *Surprise, Security, and the American Experience*. Cambridge, MA: Harvard University Press, 2004.

*Gordon, Michael R., and General Bernard E. Trainor. *The General's War: The Inside Story of the Conflict in the Gulf*. Boston: Little Brown & Co., 1991.

*Gunaratna, Rohan. *Inside al Qaeda: Global Network of Terror*. New York: The Berkley Publishing Group, 2003.

*Handel Michael I. *Masters of War: Classical Strategic Thought*, 3rd ed. London: Frank Cass Publishers, 1996.

Hersey, Paul, and Kenneth H. Blanchard. *Management of Organizational Behavior: Utilizing Human Resources*. Englewood Cliffs, NJ: Prentice-Hall, 1977.

Hoffman, Bruce. *Inside Terrorism*. New York: Columbia University Press, 1998.

*Howard, Michael. *War in European History*. Oxford: Oxford University Press, 1976.

Hunt, James G., and John D. Blair, eds. *Leadership on the Future Battlefield*. New York: Pergamon-Brassey's, 1985.

*Huntington, Samuel P. *The Soldier and the State*. New York: The Belknap Press, 1957.

*_____. *The Clash of Civilizations and the Remaking of World Order*. New York: Touchstone, 1996.

Janowitz, Morris. *The Professional Soldier: A Social and Political Portrait*. New York: Free Press, 1971.

Johnson, Kermit D. *Ethical Issues of Military Leadership*. Carlisle Barracks, PA: U.S. Army War College, 1974.

Jomini, Antoine Henri. *The Art of War*. Translated by G.H. Mendell and W.P. Craighill. 1862. Reprint, Westport, CT: Greenwood Press, 1971.

*Kagan, Donald. *The Peloponnesian War*. New York: The Penguin Press, 2003.

Kagan, Robert. *Of Paradise and Power: America and Europe in the New World Order*. New York: Vintage Press, 2004.

Kaplan, Robert. *Warrior Politics: Why Leadership Demands a Pagan Ethos*. New York: Vintage Press, 2003.

*Knox, MacGregor, and Williamson Murray. *The Dynamics of Military Revolution, 1300–2050*. New York: Cambridge University Press, 2001.

Larrabee, Eric. *Commander in Chief: Franklin Delano Roosevelt, His Lieutenants, and Their War*. New York: Harper & Row, 1987.

Lawrence, T. E. *Seven Pillars of Wisdom: A Triumph*. New York: Anchor Books, 1991.

_____. *The Arab Bulletin*. 20 August 1917. <http://www.cgsc.army.mil/carl/resources/biblio/27articles.asp>.

Lewis, Lloyd. *Sherman, Fighting Prophet*. New York: Harcourt, Brace & Co., 1932.

*Locher, James R. III. *Victory on the Potomac*. College Station, TX: Texas A&M University Press, 2004.

Luttwak, Edward N. *The Pentagon and the Art of War*. New York: Simon & Schuster, 1984.

*Macgregor, Douglas A. *Transformation Under Fire: Revolutionizing How America Fights*. Westport, CT: Praeger Publishers, 2003.

Mackey, Sandra. *Reckoning—Iraq and the Legacy of Saddam Hussein*. New York: W. W. Norton Co., 2003.

Manstein, Erich von. *Lost Victories*. Edited and translated by Anthony G. Powell. Chicago: Henry Regnery Co., 1958. Reprint, Novato, CA: Presidio Press, 1982.

McCullough, David. *Truman*. New York: Simon & Schuster, 1992.

*McMaster, H.R. *Dereliction of Duty: Lyndon Johnson, Robert McNamara, the Joint Chiefs of Staff, and the Lies That Led to Vietnam*. New York: HarperCollins Publishers, 1997.

*McPherson, James. *Battle Cry of Freedom: The Civil War Era*. New York: The Oxford University Press, 1988.

Montgomery of Alamein, Field-Marshal Viscount. *A History of Warfare*. Cleveland, OH: World Publishing Co., 1968.

*Murray, Williamson, MacGregor Knox, and Alvin Berstein. *The Making of Strategy: Rulers, States, and War*. New York: Cambridge University Press, 1994.

Musashi, Miyamoto. *A Book of Five Rings*. Woodstock, NY: The Overlook Press, 1982.

*Neustadt, Richard E., and Ernest May. *Thinking in Time*. New York: The Free Press, 1986.

Patton, George S., Jr. *War As I Knew It*. Annotated by Paul D. Harkins. Boston: Houghton Mifflin Co. 1947.

*Paret, Peter, ed. *Makers of Modern Strategy: From Machiavelli to the Nuclear Age*. Princeton, NJ: Princeton University Press, 1986.

Pogue, Forrest D. *George C. Marshall: Ordeal and Hope 1939-1942*. New York: Viking Press, 1966.

Powell, Colin. *My American Journey*. New York: Random House, 1995.

Pratt, Fletcher. *Eleven Generals, Studies in American Command*. New York: William Sloane Associates, 1949.

Ridgway, Matthew B. *Soldier: The Memoirs of Matthew B. Ridgway*. New York: Harper & Brothers, 1956.

Rommel, Erwin. *Attacks*. Vienna, VA: Athena Press, 1979.

_____. *The Rommel Papers*. Translated by Paul Findlay and edited by B. H. Liddell Hart. New York: Harcourt, Brace & Co., 1953.

Ryan, Cornelius. *A Bridge Too Far*. New York: Simon & Schuster, 1974. Reprint, New York: Popular Library, 1977.

Sarkesion, Sam C. *Beyond the Battlefield: The New Military Professionalism*. New York: Pergamon Press, 1981.

*Snider, Don, and Gayle Watkins, Project Directors. *The Future of the Army Profession*. Boston: McGraw-Hill Primis Custom Publishing, 2002.

*Stoler, Mark A. *George C. Marshall: Soldier-Statesman of the American Century*. New York: Simon & Schuster Macmillan, 1989.

Summers, Harry G., Jr. *On Strategy: The Vietnam War in Context*. Carlisle Barracks, PA: U.S. Army War College, Strategic Studies Institute, 1982.

*Sun Tzu. *The Art of War*. Translated by Samuel B. Griffith. New York: Oxford University Press, 1971.

Van Creveld, Martin L. *Command in War*. Cambridge, MA: Harvard University Press, 1985.

*_____. *Supplying War: Logistics from Wallenstein to Patton*. New York: Cambridge University Press, 1977.

Wavell, Sir Archibald P. *Soldiers and Soldiering*. New York: Avery Publishing Group, 1986.

Weigley, Russell F. *Eisenhower's Lieutenants: The Campaign of France and Germany, 1944-1945*. Bloomington, IN: Indiana University Press, 1981.

Williams, T. Harry. *McClellan, Sherman, and Grant*. New Brunswick, NJ: Rutgers University Press, 1962.

Williamson, Murray, and Major General Robert H. Scales, Jr. *The Iraq War*. Cambridge, MA: The Belknap Press of Harvard University Press, 2003.

*Winton, Harold R., and David R. Mets. *The Challenge of Change: Military Institutions and New Realities, 1918–1941*. Lincoln, NE: University of Nebraska Press, 2000.

Yildiz, Kerim. *The Kurds in Iraq: The Past, Present and Future*. London: Pluto Press, 2004.

CONCEPTUAL FOUNDATIONS

Avolio, Bruce. *Full Leadership Development: Building the Vital Forces in Organizations*. Thousand Oaks, CA: Sage Publications, 2001.

Bass, Bernard M. *Bass & Stogdill's Handbook of Leadership*. 3rd ed. New York: Free Press, 1990.

Blanchard, Kenneth H., Patricia Zigarmi, and Drea Zigarmi. *Leadership and the One Minute Manager*. New York: Morrow, 1985.

Day, David V., Stephen J. Zaccaro, and Stanley M. Halpin, eds. *Leader Development for Transforming Organizations: Growing Leaders for Tomorrow*. Mahwah, NJ: Lawrence Erlbaum Associates, 2004.

Drucker, Peter F. *The Effective Executive*. New York: Harper Collins Publishers, 2002.

Hammond, John S., Ralph L. Keeney, and Howard Raiffa. *Smart Choices: A Practical Guide to Making Better Decisions*. Boston: Harvard Business School Press, 1999.

Heifetz, Ronald. *Leadership Without Easy Answers*. Cambridge, MA: The Belknap Press of Harvard University Press, 1994.

House, Robert J., Paul J. Hanges, Mansour Javidan, Peter W. Dorfman, and Vipin Gupta, eds. *Culture, Leadership, and Organizations: The GLOBE Study of 62 Societies*. Thousand Oaks, CA: Sage Publications, 2004.

Hughes, Richard L., Robert C. Ginnett, and Gordon J. Curphy. *Leadership: Enhancing the Lessons of Experience*. New York: McGraw-Hill/Irwin, 2005.

Kotter, John P. *Leading Change*. Boston: Harvard Business School Press, 1996.

Kouzes, James M., and Barry Z. Posner. *The Leadership Challenge: How to Keep Getting Extraordinary Things Done in Organizations*. San Francisco: Jossey-Bass Publishers, 1995.

Linsky, Martin, and Ronald A. Heifetz. *Leadership on the Line: Staying Alive Through the Dangers of Leading*. Boston, MA: Harvard Business School, 2002.

Lombardi, Vince, Jr. *What It Takes To Be #1: Vince Lombardi on Leadership*. New York: McGraw-Hill, 2001.

London, Manuel. *Leadership Development: Paths to Self-Insight and Professional Growth*. Mahwah, NJ: Lawrence Erlbaum Associates, 2001.

McCann, Carol, and Ross Pigeau. *The Human in Command: Exploring the Modern Military Experience*. New York: Kluwer Academic/Plenum Publishers, 2000.

McCauley, Cynthia D., and Ellen Van Velsor, eds. *The Center for Creative Leadership Handbook of Leadership Development* 2nd ed. San Francisco: Jossey-Bass Publishers, 2004.

Murphy, Susan E., and Ronald E. Riggio, eds. *The Future of Leadership Development*. Mahwah, NJ: Lawrence Erlbaum Associates, 2003.

Northouse, Peter G. *Leadership: Theory and Practice*. Thousand Oaks, CA: Sage Publications, 2004.

Paul, Richard, and Linda Elder. *Critical Thinking: Tools for Taking Charge of Your Professional and Personal Life*. Upper Saddle River, NJ: Financial Times Prentice Hall, 2002.

Pearce, Craig L., and Jay. A. Conger. *Shared Leadership: Reframing the Hows and Whys of Leadership*. Thousand Oaks, CA: Sage Publications, 2003.

Peters, Thomas J., and Nancy Austin. *A Passion for Excellence: The Leadership Difference*. New York: Random House, 1985.

Peters, Thomas J., and Robert H. Waterman. *In Search of Excellence: Lessons from America's Best-Run Companies*. New York: HarperCollins Publishers Inc., 2004.

Salas, Eduardo, and Gary A. Klein. *Linking Expertise and Naturalistic Decision Making*. Mahwah, NJ: Lawrence Erlbaum Associates, 2001.

Schein, Edgar H. *Organizational Culture and Leadership*. New York: Jossey-Bass/John Wiley & Sons, 2004.

Waterman, Robert H., and Thomas J. Peters. *In Search of Excellence*. New York: Harper & Row, 1982.

Yukl, Gary A. *Leadership in Organizations* 6th ed. Upper Saddle River, NJ: Prentice-Hall, 2005.

Zaccaro, Stephen J. *The Nature of Executive Leadership: A Conceptual and Empirical Analysis of Success*. Washington, DC: American Psychological Association, 2001.

CULTURE AND REGIONAL STUDIES

Andric, Ivo. *The Bridge on the Drina (Phoenix Fiction Series)*.Chicago: University of Chicago Press, 1977.

Armstrong, Karen. *Islam: A Short History*. New York: Modern Library; 2000.

Ayittey, George B.N. *Africa in Chaos: A Comparative History*. New York: Palgrave Macmillan, 1998.

Bahmanyar, Mir, and Ian Palmer. *Afghanistan Cave Complexes, 1979-2004: Mountain Strongholds of the Mujahideen, Taliban & Al Qaeda (Fortress)*. Oxford, UK: Osprey Publishing, 2004.

Baker, Peter, and Susan Glasser. *Kremlin Rising: Vladimir Putin's Russia and the End of Revolution*. New York: A Lisa Drew Book/Scribner, 2005.

Baker, William G. *The Cultural Heritage of Arabs, Islam and the Middle East*. Dallas, TX: Brown Books Publishing Group, 2003.

Barzini, Luigi. *The Europeans*. New York: Penguin (Non-Classics), 1984.

Cohen, Stephen P. *The Idea of Pakistan*. Washington, DC: Brookings Institution Press, 2004.

Cordesman, Anthony H. *Lessons of Afghanistan: War Fighting, Intelligence, and Force Transformation* (Significant Issues Series, Vol. 24, No. 4). Washington, DC: Center for Strategic & International Studies, 2002.

Duus, Peter. *Modern Japan*. Boston, MA: Houghton Mifflin Company, 1998.

Esposito, John L. *Islam: The Straight Path*. New York: Oxford University Press, 1998.

_____. *Unholy War: Terror in the Name of Islam*. New York: Oxford University Press, 2002.

_____. *What Everyone Needs to Know About Islam*. New York: Oxford University Press, 2002.

Fairbank, John King, and Merle Goldman. *China: A New History*. Cambridge, MA: The Belknap Press of Harvard University Press, 1998.

Friedman, Thomas. *From Beirut to Jerusalem* Revised Edition. New York: Farrar, Straus and Giroux, 1991.

Fuller, Graham E. *The Future of Political Islam*. New York: Palgrave Macmillan, 2003.

Girardet, Edward, Jonathan Walter, and Charles Norchi, eds. *Afghanistan: Crosslines Essential Field Guides to Humanitarian and Conflict Zones* 2nd ed. Versoix, Switzerland: Media Action International, 2004.

Goldschmidt, Arthur, Jr. *A Concise History of the Middle East* 7th ed.Boulder, CO: Westview Press, 2001.

Graff, David A., and Robin Higham *A Military History of China*. Boulder, CO: Westview Press, 2002.

Graham, Richard. *The Idea of Race in Latin America: 1870-1940*. Austin, TX: University of Texas Press, 1990.

Green, Michael J. *Arming Japan: Defense Production, Alliance Politics, and the Postwar Search for Autonomy*. New York: Columbia University Press, 1995.

Harrison, Lawrence E., and Samuel P. Huntington, eds. *Culture Matters: How Values Shape Human Progress*. New York: Basic Books/Perseus Books Group, 2001.

Heidhues, Mary F. Somers. *Southeast Asia: A Concise History*. New York: Thames & Hudson, Inc., 2000.

Iliffe, John. *Africans: The History of a Continent*. Cambridge, UK: Cambridge University Press, 1995.

Joseph, Richard A. *State, Conflict, and Democracy in Africa*. Boulder, CO: Lynne Rienner Publishers, 1998.

Kaplan, Robert. *Eastward to Tartary: Travels in the Balkans, the Middle East, and the Caucasus*. New York: Random House, 2000.

_____. *Soldiers of God: With Islamic Warriors in Afghanistan and Pakistan*. New York: Vintage Books, 2001.

Karnow, Stanley. *In Our Image: America's Empire in the Philippines*. New York: Ballantine Books, 1990.

_____. *Vietnam: A History*. 2nd ed. New York: Penguin Books, 1997.

Landes, David S. *The Wealth and Poverty of Nations: Why Some Are So Rich and Some So Poor*. New York: W.W. Norton & Company, Inc., 1999.

Lebra, Takie Sugiyama. *Japanese Patterns of Behavior*. Honolulu, HI: University of Hawaii Press, 1976.

Leinbach, Thomas R., and Richard Ulack. *Southeast Asia: Diversity and Development*. Upper Saddle River, NJ: Prentiss Hall, 1999.

Liberthal, Kenneth. *Governing China: From Revolution Through Reform*. New York: W.W. Norton & Company, Inc., 1995

Loveman, Brian. *The Politics of Anti-Politics: The Military in Latin America*. Lanham, MD: SR Books, 1997.

Ludden, David. *India and South Asia: A Short History*. Oxford, UK: Oneworld Publications, 2002.

Mecham, J. Lloyd. *Church and State in Latin America: A History of Politico-Ecclesiastical Relations*. Chapel Hill, NC: The University of North Carolina Press, 1969.

Mote, Victor L. *Siberia: Worlds Apart*. Boulder, CO: Westview Press, 1998

Nakash, Yitzhak. *Shi'is of Iraq*. Princeton, NJ: Princeton University Press, 2003.

Nydell, Margaret. *Understanding Arabs—A Guide for Modern Times*. Boston, MA: Intercultural Press, 2005.

Oberdorfer, Don. *The Two Koreas: A Contemporary History*. New York: Basic Books/Perseus Books Group, 2001.

Oliker, Olga, and Thomas S. Szayna. *Faultlines of Conflict in Central Asia and the South Caucasus: Implications for the U.S. Army*. Santa Monica, CA: RAND Corporation, 2003.

Owen, Norman G., David Chandler, and William R. Roff eds. *The Emergence of Modern Southeast Asia: A New History*. Honolulu, HI: University of Hawaii Press, 2004.

Patai, Raphael. *Arab Mind*. New York: Scribner, 1983. Revised Edition, Long Island City, NY: Hatherleigh Press, 2002.

Pye, Lucian W., and Mary W. Pye. *Asian Power and Politics: The Cultural Dimensions of Authority*. Cambridge, MA: The Belknap Press of Harvard University Press, 1988.

Rashid, Ahmed. *Taliban: Militant Islam, Oil and Fundamentalism in Central Asia*. New Haven, CT: Yale University Press, 2001.

Reader, John. *Africa: A Biography of a Continent*. New York: Alfred A. Knopf, Inc., 1998.

Richmond, Yale. *From Da to Yes: Understanding the East Europeans*. Yarmouth, ME: Intercultural Press, 1995.

Shambaugh, David. *Modernizing China's Military: Progress, Problems, and Prospects*. Los Angeles, CA: University of California Press, 2002.

Skidmore, Thomas E., and Peter H. Smith. *Modern Latin America, Sixth Edition*. New York: Oxford University Press, Inc., 2001.

Stern, Jessica. *Pakistan's Drift into Extremism: Allah, the Army, and America's War on Terror*. Armonk, NY: M.E. Sharpe, Inc., 2004.

Storti, Craig. *Old World/New World: Bridging Cultural Differences—Britain, France, Germany and the U.S*. Boston, MA: Intercultural Press, 2001.

Talbott, Strobe. *Engaging India: Diplomacy, Democracy, and the Bomb*. Washington, DC: Brookings Institution Press, 2006.

Terrill, Ross. *The New Chinese Empire: And What It Means for the United States*. New York: Basic Books/Perseus Books Group, 2003.

Winn, Peter. *Americas: The Changing Face of Latin America and the Caribbean*. Los Angeles, CA: University of California Press, 1992.

Yamada, Haru, and Deborah Tannen. *Different Games, Different Rules: Why Americans and Japanese Misunderstand Each Other*. New York: Oxford University Press, 2002.

Index

Entries are by paragraph number unless stated otherwise. Leaders quoted or highlighted in vignettes appear in bold.

Entries are by paragraph number unless stated otherwise

C

Entries are by paragraph number unless stated otherwise.

Entries are by paragraph number unless stated otherwise

Entries are by paragraph number unless stated otherwise.

J-K

L

Entries are by paragraph number unless stated otherwise

Entries are by paragraph number unless stated otherwise.

Entries are by paragraph number unless stated otherwise

FM 6-22
12 October 2006

By Order of the Secretary of the Army:

PETER J. SCHOOMAKER
General, United States Army
Chief of Staff

Official:

JOYCE E. MORROW
Administrative Assistant to the
Secretary of the Army

0626402

Distribution: *Active Army, Army National Guard, and US Army Reserve:* To be distributed in accordance with initial distribution number 110180, requirements for FM 6-22.